A-Z
ELECTROMAGNETISM

A-Z ELECTROMAGNETISM

Prof. G.K. Bose

CENTRUM PRESS
NEW DELHI-110002 (INDIA)

CENTRUM PRESS

H.O.: 4360/4, Ansari Road, Daryaganj,
New Delhi-110 002 (India)
Ph.: 23278000, 23261597

B.O.: No. 1015, Ist Main Road, BSK IIIrd Stage
IIIrd Phase, IIIrd Block,
Bangalore - 560 085 (India)
Tel.: 080-41723429
Visit us at: www.centrumpress.com

A-Z Electromagnetism

First Edition, 2009

ISBN 978-93-80106-17-5

PRINTED IN INDIA

Printed at Salasar Imaging Systems, Delhi-110035 (India)

Contents

Preface

Electromagnetic Field Theory is one of the fundamental courses that an electrical and computer engineering student is required to take in order to gain a physical understanding of the foundations and the heritage of the field that will occupy his or her professional life for the several decades following graduation. The acquiring of an appreciation for the laws of nature that govern and limit the speed of the smallest computer chip continue to be crucial as this speed approaches the ultimate limit. With the many changes that are occurring in undergraduate curriculums due to the rapid development of new technologies and hence additional courses, it is common to find that only one course in electromagnetic theory is now required for students.

The environment is now thoroughly polluted by man-made sources of electromagnetic radiation with frequencies and magnitudes never before present. Man's activities have probably changed the earth's electromagnetic background to a greater degree than they have changed any other natural physical attribute of the earth- whether the land, water, or atmosphere. The evidence now indicates that the present abnormal electromagnetic environment can constitute a health risk.This book is our attempt to synthesize the various aspects of the role of electricity and magnetism in physics, and to emphasize their underlying unity.

Author

Chapter 1

Introduction to Electromagnetism

Electromagnetism is the physics of the electromagnetic field: a field which exerts a force on particles that possess the property of electric charge, and is in turn affected by the presence and motion of those particles.A changing magnetic field produces an electric field (this is the phenomenon of electromagnetic induction, the basis of operation for electrical generators, induction motors, and transformers).

Similarly, a changing electric field generates a magnetic field. Because of this interdependence of the electric and magnetic fields, it makes sense to consider them as a single coherent entity–the electromagnetic field. The magnetic field is produced by the motion of electric charges, *i.e.*, electric current. The magnetic field causes the magnetic force associated with magnets.

Electricity and Magnetism (or ε & *M*, as they are known in the trade) are "weird" phenomena because the palpable *forces* they generate on objects seem to come from nowhere–nothing is "touching" the objects and yet they are moved. The related fact that we are unable to wilfully exert significant electrical or magnetic forces directly on objects around us using any combination of muscles or mechanical devices removes ε & *M* still further from our personal sensory experience and thus makes them seem "weirder."

Even the most seasoned ε & *M* veteran still experiences a sense of primitive wonder when a magnet on top of the table moves "by magic" under the influence of another magnet underneath the table.

On the one hand, this makes ε & *M* a fun subject to study.

On the other hand, it makes ε & *M* hard to teach, because it will never make "common sense" like nuts-and-bolts Mechanics. *C'est la vie.* As our first foray into "Weird Science" it is only fitting that ε & *M* should be something we know is there but that we will just have to get used to instead of ever hoping to rectify it with our common sense. It is, of course, "common sense" itself that is defective.

DIRECT FORCE LAWS

There are two fundamental kinds of forces in ε & *M*: the *electrostatic* force between two *charges* and the *magnetic* force between two *currents*.

- The Electrostatic Force
- The Magnetic Force

The Electrostatic Force

First, what is a *charge*? We don't know! But then, we don't know what a *mass* is, either, except in terms of its behaviour: a mass resists acceleration by forces and attracts other masses with a gravitational force. The analogy is apt, in the sense that electrical charges exert forces on each other in almost exactly the same way as masses do, except for two minor differences. Recall Newton's universal law of gravitation in its most democratic form: the force $\vec{F}_{12}^{G}$ acting *on* body (mass m_2) *due to* body (mass m_1) is

$$\vec{F}_{12}^{G} = -G\frac{m_1 m_2}{r_{12}^2}\hat{r}_{12}$$

where G is the Universal Gravitational Constant, r_{12} is the distance between the two masses and $\hat{r}_{12}$ is the unit vector pointing *from* #1 to #2. The *electrostatic* force $\vec{F}_{12}^{E}$ between two *charges* q_1 and q_2 is of exactly the same form:

$$\vec{F}_{12}^{E} = k_E\frac{q_1 q_2}{r_{12}^2}\hat{r}_{12}$$

where k_E is some constant to make all the units come out right.

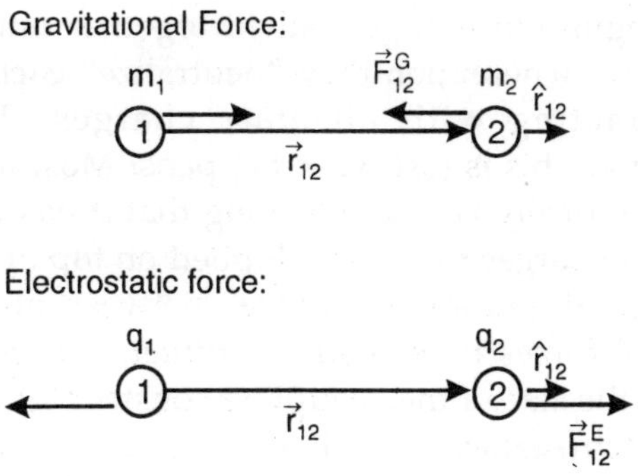

Fig. Comparison of the Gravitational Force $\vec{F}^G_{12}$ on Mass m_2 due to Mass m_1 and the Electrostatic (Coulomb) Force $\vec{F}^E_{12}$ on Electric Charge q_2 due to Charge q_1.

This force law, also known as the COULOMB FORCE, has almost the same qualitative earmarks as the force of gravity: the force is "central"–*i.e.,* it acts along the line joining the centres of the charges–and drops off as the inverse square of the distance between them; it is also proportional to each of the charges involved.

So what are the "minor differences?" Well, the first one is in the *sign*. Both "coupling constants" (G and k_E) are defined to be *positive;* therefore the–sign in the first equation tells us that the gravitational force $\vec{F}^G_{12}$ on mass #2 is in the *opposite direction* from the unit vector $\hat{r}_{12}$ pointing form #1 to #2—*i.e.,* the force is *attractive,* back toward the other body. All masses attract all other masses gravitationally; there are no repulsive forces in gravity.

Another way of putting it would be to say that "there are no negative masses." By contrast, electric charges come in both positive (+) and negative (–) varieties; moreover, Equation tells us that the electrical force $\vec{F}^E_{12}$ on charge #2 is in the *same* direction as $\hat{r}_{12}$ as long as the product $q_1\ q_2$ is positive–*i.e.,* charges of *like* sign [*both* + or *both* –] *repel* whereas *unlike* charges *attract.*

This means that a positive charge and a negative charge

of equal magnitude will get pulled together until their net charge is zero, whereupon they "neutralize" each other and cease interacting with all *other* charges. To a good approximation, this is just what happens! Most macroscopic matter is electrically *neutral,* meaning that it has the positive and negative charges pretty much piled on top of each other.

The second "minor difference" between electrical and gravitational forces is in their *magnitudes*. Of course, each depends on the size of the "coupling constant" [G for gravity *vs.* k_E for electrostatics] as well as the sizes of the "sources" [m_1 and m_2 for gravity *vs.* q_1 and q_2 for electrostatics] so any of magnitude has to be in reference to "typical" examples. Let's choose the heaviest stable elementary particle that has both charge and mass: the *proton,* which constitutes the nucleus of a hydrogen atom. A proton has a positive charge of

$$\varepsilon = 1.60217733(49) \times 10^{-19} \text{ coulomb (abbreviated C)}$$

and

$$m_p = 1.6726231(10) \times 10^{-27} \text{ kg}$$

For *any* separation distance *r*, two protons *attract* each other (gravitationally) with a force whose magnitude F_G is $\dfrac{Gm_r^2}{k_E e^2}$ times the magnitude F_E of the (electrostatic) force with which they *repel* each other. This ratio has an astonishing value of 0.80915×10^{-36}–the gravitational attraction between the two protons is roughly a *trillion trillion trillion* times *weaker* than the electrostatic repulsion. The electrical force wins, hands down. However, *in spite of its phenomenal puniness, gravity can overcome all other forces if enough mass gets piled up in one place.*

The Magnetic Force

"Laws" of ε and M are so symmetric between electrical and magnetic phenomena that most Physicists are extremely frustrated by the fact that no one has ever been able to conclusively demonstrate the existence (other than theoretical) of a "magnetic charge" (also known as a *magnetic monopole*). If there were magnetic charges, the magnetic force equation would look just like the gravitational and electrostatic force laws above and this part of the description would be nice and simple.

Alas, this is not the case. Static (constant in time) magnetic phenomena are generated instead by the steady *motion* of electric charges, represented by a *current I* (the charge passing some fixed point per unit time) in some direction $\vec{\ell}$. Usually (at least at the outset) we talk about currents flowing in a *conductor* (*e.g.* a wire) through which the charges are free to move with minimal resistance. Then $\vec{\ell}$ is a vector length pointing along the wire, or (if the wire is curved) $d\vec{\ell}$ is an infinitesimal *element* of the wire at some point. We may then think in terms of a "current element" $Id\vec{\ell}$.

One such current element $I_1 d\vec{\ell}_1$ exerts a magnetic forced $\vec{F}_{12}^M$ on a second current element $I_2 d\vec{\ell}_2$ at a distance $\hat{r}_{12}$ (the vector from #1 to #2) given by

$$d\vec{F}_{12}^M = k_M \frac{I_1 I_2}{r_{12}^2} d\vec{\ell}_2 \times (d\ell_1 \times \hat{r}_{12})$$

where k_M is yet another unspecified constant to make all the units come out right and again $\hat{r}_{12}$ is the *unit vector* defining the direction *from #1 to #2*.

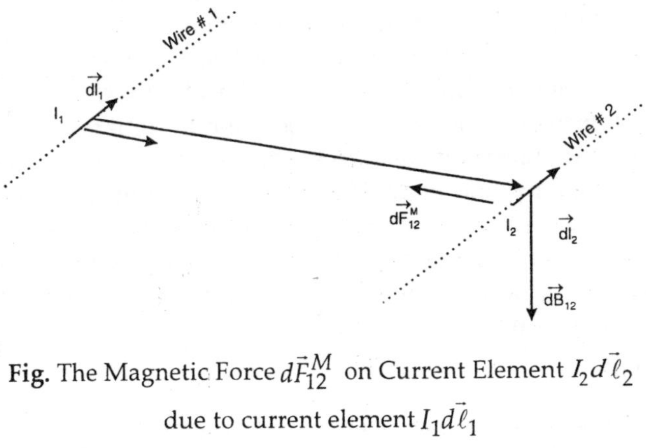

Fig. The Magnetic Force $d\vec{F}_{12}^M$ on Current Element $I_2 d\vec{\ell}_2$ due to current element $I_1 d\vec{\ell}_1$

This ugly equation does give us some important

qualitative hints about the force between two current-carrying wires: the force between any two *elements* of wire drops off as the inverse square of the distance between them, just like the gravitational and electrostatic forces although this isn't much use in guessing the force between real current-carrying wires, which don't come in infinitesimal lengths and the force is in a direction perpendicular to both wires.

The magnetic forces will act between two *parallel* wires: we can visualize a distance vector $\vec{r}$ from the first wire over to the second wire; let it be perpendicular to both for convenience. The "Right-hand Rule" will then tell us the *direction* of $(d\vec{\ell}_1 \times \hat{r}_{12})$: if we "turn the screw" in the sense of cranking through the angle *from* $d\vec{\ell}_1$ *to* $\hat{r}_{12}$, a right-handed screw [the conventional kind] would move in the direction labeled $d\vec{B}_{12}$.

This is the direction of $(d\vec{\ell}_1 \times \hat{r}_{12})$. Now if we crank $d\vec{\ell}_2$ into $d\vec{B}_{12}$, the turn of the screw will cause it to head back toward the first wire. Seriously, *no one* is particularly enthused over this equation! All anyone really retains from this intricate exercise is the following pair of useful rules:

- Two parallel wires with electrical currents flowing in the *same* direction will *attract* each other.
- Two parallel wires with electrical currents flowing in *opposite* directions will *repel* each other.

Nevertheless, electrical engineers and designers of electric motors and generators need to know just what sorts of forces are exerted by one complicated arrangement of current-carrying wires on another; moreover, once it had been discovered that moving charges create this weird sort of action-at-a-distance, no one wanted to just give up in disgust and walk away from it.

FIELDS

In Classical Mechanics we found several conceptual aids that not only made calculations easier by skipping over inessential details but also made it possible to carry around the bare essence of Mechanics in our heads in a small number of compact "Laws." This is generally regarded as a good thing, although of course we pay a price for every entrenched paradigm–we may lose the ability to without filtering our experience through models.

Debate to the philosophers, psychologists and mystics; it is true even in Physics, however, that the more successful the paradigm the bigger the blind spot it creates for alternative descriptions of the same phenomena. This bothers most Physicists, too, but there doesn't seem to be a practical alternative; so we content ourselves with maintaining an awareness of our own systematic prejudices.

Perhaps the best example of this from the days of "Classical" Physics [*i.e.*, before Relativity and Quantum Mechanics rained confusion down on all of us] is the invention of the Electric and Magnetic Fields, written $\vec{E}$ and $\vec{B}$, respectively.

- First, calculate the field due to the *source* charge or current.
- Then calculate the *force* on a *test* charge or current *due to* that field.

The Electric Field

The electric field $\vec{E}$ at any point in space is defined to be the *force per unit test charge* due to all the other charges in the universe. That is, there is probably no "test charge" q there to experience any force, but *if there were* it would experience a force

$$\vec{F}_E = q\vec{E}$$

Note that since the force is a vector, $\vec{E}$ is a *vector field*.

Since by definition $\vec{E}$ is there even if there *isn't* any test charge present, it follows that there is an electric field at every point in space, all the time.

Is the Electric Field real? No. Yes. You decide. This paradigm makes everything so much easier that most

Physicists can't imagine thinking about ε & *M* any other way. Does this blind us to other possibilities? Undoubtedly.

A single isolated electric "source" charge *Q* generates a *spherically symmetric* electric field at any point in space specified by the vector distance $\vec{r}$ from *Q* to that point.

$$\vec{E} = k_E \frac{Q}{r^2} \hat{r}$$

That is, the field $\vec{E}$ is *radial* in the direction of the radius vector and has the same *magnitude E* at all points on an imaginary spherical surface a distance *r* from *Q*.

It might be helpful to picture the *acceleration of gravity* as a similar *vector field*:

$$\vec{g} = -G \frac{M_E}{r^2} \hat{r}$$

i.e. $\vec{g}$ always points back toward the centre of the Earth (mass M_E) and drops off as the inverse square of the distance *r* from the centre of the Earth.

The Magnetic Field

Any current element $Id\vec{\ell}$ contributes $d\vec{B}$ to the magnetic field $\vec{B}$ at a given point in space:

$$d\vec{B} = k_M \frac{Id\vec{\ell} \times \hat{r}}{r^2}$$

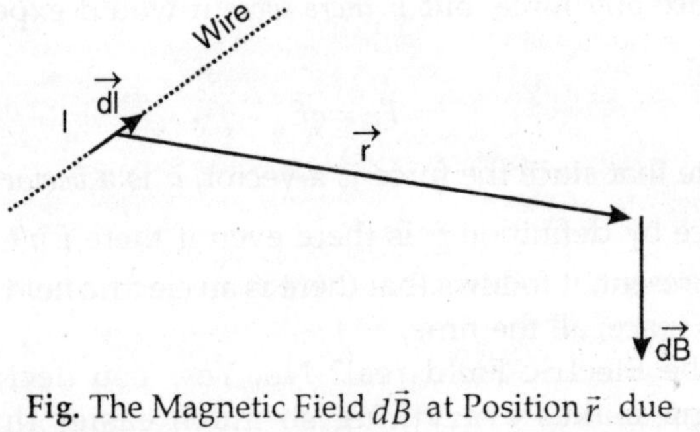

Fig. The Magnetic Field $d\vec{B}$ at Position $\vec{r}$ due to a Current Element $Id\vec{\ell}$ at the Origin

where $\hat{r}$ is the unit vector in the direction of $\vec{r}$, the vector distance from the current element to the point in space where the magnetic field is being evaluated. Eq. is known as the Law of Biot and Savart. It is still not perfectly transparent, I'm sure you will agree, but it beats Equation.

Superposition

While it may seem obvious, it bears saying that the electric fields due to several different "source" charges or the magnetic fields due to several different "source" current elements are just added together (vectorially, of course) to make the net $\vec{E}$ or $\vec{B}$ field. Horrible as it might seem, this might in principle *not* be true–we might have to "add up" such fields in some hopelessly more complicated way.

THE LORENTZ FORCE

We can now put the second part of the procedure calculating the *forces* on a test charge due to known Fields into a very compact form combining both the electric and the magnetic forces into one equation. If a particle with charge q and mass m moves with velocity $\vec{v}$ in the combination of a uniform electric field $\vec{E}$ and a uniform magnetic field $\vec{B}$, the net force acting on the particle is the Lorentz Force, which can be written (in one set of units)

$$\vec{F} = q\left(\vec{E} + \frac{\vec{v}}{c} + \vec{B}\right),$$

where (for now) we can think of c as just some constant with units of velocity.

If $\vec{E} = 0$ and $\vec{v}$ is perpendicular to $\vec{B}$, the Lorentz force is perpendicular to both $\vec{B}$ and the momentum $\vec{p} = m\vec{v}$. The force will deflect the momentum sideways, changing its direction but not its magnitude.

As $\vec{p}$ changes direction, $\vec{F}$ changes with it to remain ever perpendicular to the velocity–this is an automatic property of the cross product–and eventually the *orbit* of the particle closes back on itself to form a circle. In this way the magnetic field produces Uniform Circular Motion with the plane of the circle perpendicular to both $\vec{v}$ and $\vec{B}$.

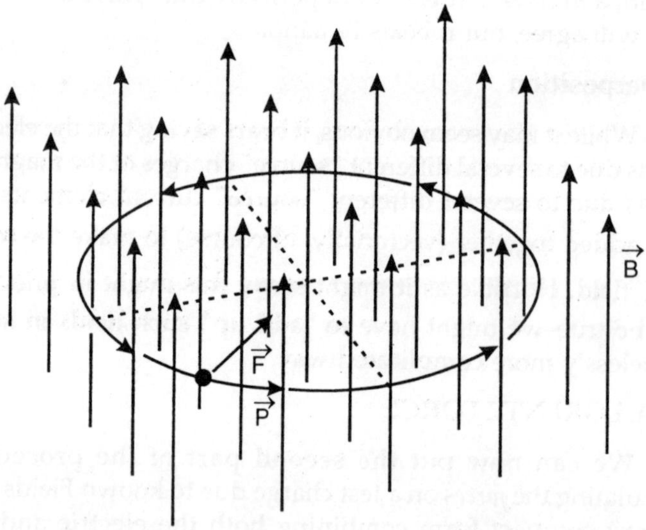

Fig. Path of a Charged Particle with Momentum $\vec{p}$ in a Uniform, Static Magnetic Field $\vec{B}$ Perpendicular to $\vec{p}$

Using Newton's Second Law and a general knowledge of circular motion, one can derive a formula for the *radius* of the circle (r) in terms of the *momentum* of the particle ($p = mv$), its *charge* (q) and the magnitude of the *magnetic field* (B). In "Gaussian units" (grams, centimeters, Gauss) the formula reads.

$$\tau = \frac{pc}{qB}.$$

It is also interesting to picture qualitatively what will happen to the particle if an *electric* field $\vec{E}$ is then applied *parallel* to $\vec{B}$: since $\vec{E}$ accelerates the charge in the direction of $\vec{E}$, which is also the direction of $\vec{B}$, and since $\vec{B}$ only produces a force when the particle moves *perpendicular* to $\vec{B}$, in effect the "perpendicular part of the motion" is unchanged (circular motion) while the "parallel part" is unrestricted acceleration. The path in space followed by the particle will be a spiral with steadily increasing "pitch":

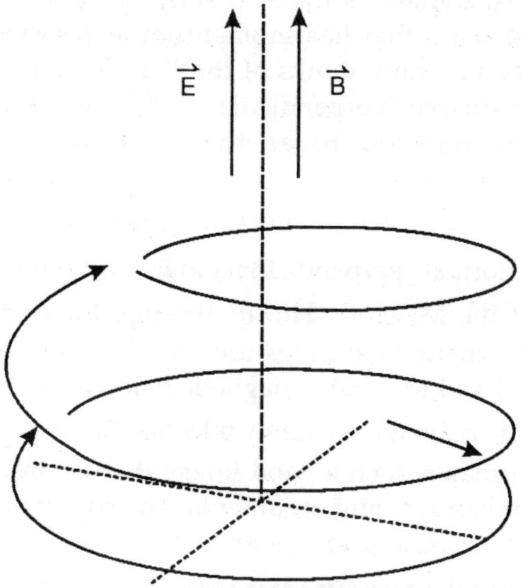

Fig. Path of a Charged Particle in Parallel $\vec{E}$ and $\vec{B}$ Fields

Field Lines and Flux

The uniform magnetic field is pictured as a forest of little parallel arrows of equal length, equally spaced. Something like this is always necessary if we want to make a visual representation of $\vec{B}$, but it leaves a lot to be desired. For instance, a uniform magnetic field has the same magnitude and direction at *every* point in space, not just where the lines are drawn. The magnetic *force,* if any, is *never* in the direction of the "lines of $\vec{B}$" but rather perpendicular to them. Nevertheless, the visual appeal of such a graphical representation in terms of "field lines" is so compelling that a whole description of ε & *M* has been developed in terms of them. In that description one speaks of "lines per unit area" as a measure of the *strength* of an electric or magnetic field. The analogy is with *hydrodynamics,* the flow of incompressible fluids "lines" of fluid flow if we drop packets of dye in the water.

In fluid dynamics there is actually "stuff" flowing, a transfer of mass that has momentum and density. In that context one naturally thinks of the Flux of material through imaginary surfaces perpendicular to the flow and indeed is sometimes referred to as the *magnetic flux per unit* (perpendicular) *area.*

By the same token, if "lines" of $\vec{B}$ pass through a surface of area A normal (perpendicular) to $\vec{B}$, then we can (and do) talk about the Magnetic Flux Φ through the surface; Φ has units of magnetic field times area. If we want, we can turn this around and say that a magnetic field has units of *flux per unit area.* Even though we rarely take this "lines of $\vec{B}$" business literally, it makes such a good image that we make constant use of it in handwaving arguments. The concept of magnetic flux is well ensconced in modern ε & M terminology.

POTENTIALS AND GRADIENTS

Recall from Mechanics that if we move a particle a vector distance $d\vec{\ell}$ under the influence of a force $\vec{F}$, that force does $dW = \vec{F}.d\vec{\ell}$ worth of *work* on the particle–which appears as *kinetic energy. Etc.* If the force is due to the action of an electric field $\vec{E}$ on a charge q, the work done is $dW = q\vec{E}.d\vec{\ell}$. This work gets "stored up" as *potential energy* V as usual: $dV = -dW$. Just as we defined $\vec{E}$ as the *force per unit charge,* we now define the Electric Potential ϕ to be the *potential energy per unit charge, viz.*

$$dV = qd\phi \text{ where } d\phi = -\vec{E}.d\vec{\ell}$$

or, summing the contributions from all the infinitesimal elements $\vec{\ell}$ of a finite path through space in the presence of electric fields,

$$\phi \equiv -\int \vec{E}.d\vec{\ell}$$

When multiplied by q, ϕ gives the potential *energy* of the charge q in the electric field $\vec{E}$.

Just as we quickly adapted our formulation of Mechanics to use *energy* (potential and kinetic) as a starting point instead of *force*, in ε & *M* we usually find it easier to start from $\phi(\vec{r})$ as a function of position $(\vec{r})$ and derive $\vec{E}$ the same way we did in Mechanics:

$$\vec{E} \equiv -\vec{\nabla}\phi$$

where, as before,

$$\vec{\nabla} \equiv \hat{x}\frac{\partial}{\partial x} + \hat{y}\frac{\partial}{\partial y} + \hat{z}\frac{\partial}{\partial z}$$

The most important example is, of course, the electric potential due to a single "point charge" Q at the origin:

$$\phi(\vec{r}) = k_E \frac{Q}{r}$$

Note that $\phi(r) \to 0$ as $r \to 0$. This is a convenient convention.

$$\vec{\nabla}\left(\frac{1}{r}\right) = -\frac{\hat{r}}{r^2}.$$

Electric potential is most commonly measured in *volts* (abbreviated *V*) after Count Volta, who made the first useful batteries. We often speak of the "voltage" of a battery or an appliance. The latter does not ordinarily have any electric potential of its own, but it is designed to be *powered* by a certain "voltage." A light bulb would be a typical case in point. The *volt* is actually such a familiar unit that eletric *field* is usually measured in the derivative unit, *volts per meter* (*V/m*). The electrostatic potential ϕ is often referred to as the Scalar Potential, which immediately suggests that there must be such a thing as a vector potential too. Just so. The vector potential $\vec{A}$ is used to calculate the *magnetic* field $\vec{B}$ but not quite as simply as we get $\vec{E}$ from $\vec{\nabla}\phi$. In this case we have to take the "curl" of $\vec{A}$ to get $\vec{B}$:

$$\vec{B} = \vec{\nabla} \times \vec{A} \cdot$$

UNITS

When Physicists are working out problems "formally" (that is, trying to understand "how things behave") they are usually only concerned with deriving a formula which describes the behaviour, not so much with getting "numbers" out of the formula. This is why we can tolerate so much confusion in the details of the alternate electromagnetic unit systems.

We never actually calculate any "answers" that an engineer could use to build devices with; we simply derive a formula for such calculations, preferably in a form as free of specific units as possible, and leave the practical details up to the engineer. As long as electric and magnetic fields are not both involved in the same problem, one can usually stick to familiar units expressed in a reasonably clear fashion.

Electrical Units

The old-fashioned version of this saga, in which one picks either Volts or Coulombs as the "fundamental" unit and derives the rest from that. Today the Ampere [A] is actually the most basic unit; it is *defined* to be the current required to flow in *both* of two "very long" parallel wires 1 *m* apart in order to give a *magnetic force per unit length* of exactly 2 × 10^{-7} *N*/*m* acting on each wire. No, I'm not kidding. Then the Coulomb [C] is defined as the electric charge that flows past any point in 1 *s* when a steady current of 1 A is maintained in a wire. *I.e.* we have 1 C = 1 A-*s*.

Coulombs and Volts

As electric *charge* is usually measured in COULOMBS (abbreviated C). If we take this as a fundamental unit, we can analyse the definition of the *volt* (V): moving a charge of q = 1 C through an electric potential difference $\Delta\phi$ = 1V produces a potential energy difference of $\Delta V = 1$ J. Therefore a volt is a joule per coulomb.

If we prefer to think of the *volt* as a more fundamental unit, we can turn this around and say that a Coulomb is a joule per volt.

Electron Volts

We can also take advantage of the fact that Nature supplies electric charges in integer multiples of a fixed quantity of charge to define some more "natural" units. For instance, the electric charge of an electron is $-e$ [where e is the charge of a proton]. An Electron Volt (eV) is the kinetic *energy* gained by an electron [or any other particle with the same size charge] when it is accelerated through a one volt (1 V) electric potential. Moving a charge of 1 C through a potential of 1 V takes 1 J of work (and will produce 1 J of kinetic energy), so we know immediately from Equation that

$$1 \text{ eV} = 1.60217733(49) \times 10^{-19} \text{ J}$$

This is not much energy if you are a toaster, but for an electron (which is an *incredibly tiny* particle) it is enough to get it up to a velocity of 419.3828 km/s, which is 0.14% of the speed of light! Another way of looking at it is to recall that we can express *temperature* in energy units using Boltzmann's constant as a conversion factor. You can easily show for yourself that 1 eV is equivalent to a temperature of 11,604 degrees Kelvin or about 11,331C. So in the microscopic world of electrons the eV is a pretty convenient (or "natural") unit. But not in the world of toasters and light bulbs. So let's get back to "conventional" units.

Amperes

Electric currents (the rate at which charges pass a fixed point in a wire, for instance) have dimensions of *charge per unit time.* If the Coulomb is our chosen unit for electric charge and we retain our fondness for *seconds* as a time unit, then *current* must be measured in *coulombs per second.* We call these units Amperes or Amps [abbreviated A] after a Frenchman named Ampère.

1A [Ampere] = 1 C/s [Coulomb per second]

It makes about as much sense to give the coulomb per second its own name as it would to make up a name for meters per second. No one frets over the complexity of expressing speed in m/s or kph or whatever –in fact it serves as a good reminder that velocity is a rate of change of distance with time-but for some reason we feel obliged to give C/s their own name. Ah well, it is probably because all this electrical stuff is so weird.

The Coupling Constant

We are now ready to define our electrical "coupling constant" k_E. Referring to Equation we have;

$$\phi[V] = k_E \frac{Q[C]}{r[m]}$$

which we can rearrange to read.

Thus k_E must have *dimensions* of {electric potential times distance per unit charge}; we can pick *units* of *V*-m/C to stick with convention. This still doesn't tell us the *value* of k_E. This must be *measured*. The result is

$$k_E = 8.98755 \times 10^9 \text{ V-m/C}$$

Gauss vs. Tesla

There are two "accepted" units for the magnetic field $\vec{B}$: Gauss [abbreviated G] and TESLA [abbreviated T]. Needless to say, both are named after great ε & *M* researchers. The former is handy when describing *weak* magnetic fields--for instance, the Earth's magnetic field is on the order of 1 G–but the unit that goes best with our selected electrical units (because it is defined in terms of meters and coulombs and seconds) is the TESLA. Fortunately the conversion factor is simple:

$$1 \text{ T} = 10{,}000 \text{ G.}$$

The TESLA is also defined in terms of the WEBER [W] (named after guess whom), a conventional unit of magnetic *flux*. The definition is

$$1 \text{ T} \equiv 1 \text{ W/m}^2 \text{ or } 1\text{W} = 1 \text{ T} \times 1 \text{ m}^2$$

If you're interested. So referring back to Equation, we have

$$B[\text{Tesla}) = k_M \frac{I[A]d\ell[m]}{(r[m])^2}$$

Which we can rearrange to read

$$k_M = \frac{B[TESLA](r[m])^2}{I[A]d\ell[m]}$$

So that k_M must have *dimensions* of magnetic flux per unit current per unit length or *units* of W/A-m. Its *value* is again determined by experiment:

$$k_M = 10^7 \text{ W/A-m}$$

MAGNETIC FIELD

The high permeability of individual iron filings causes the magnetic field to be larger at the ends of the filings. This causes individual filings to attract each other, forming elongated clusters that trace out the appearance of lines. It would not be expected that these "lines" be precisely accurate field lines for this magnet; rather, the magnetization of the iron itself would be expected to alter the field somewhat. In physics, a *magnetic field* is a vector field that permeates space and which can exert a *magnetic force* on moving electric charges and on magnetic dipoles (such as permanent magnets).

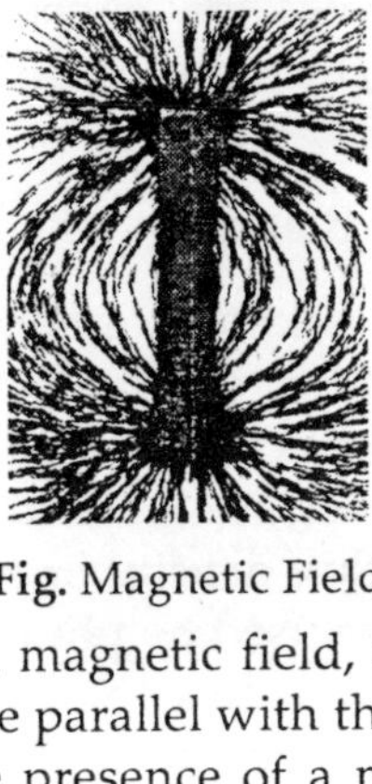

Fig. Magnetic Field

When placed in a magnetic field, magnetic dipoles tend to align their axes to be parallel with the magnetic field when iron filings are in the presence of a magnet. In addition, a changing magnetic field can induce an electric field.

Magnetic fields surround and are created by electric currents, magnetic dipoles, and changing electric fields. Magnetic fields also have their own energy, with an energy density proportional to the square of the field intensity.

The electric field and the magnetic field are tightly interlinked, in two senses. First, changes in either of these fields can cause changes in the other, according to Maxwell's equations. Second, according to Einstein's theory of special relativity, a magnetic force in one inertial frame of reference may be an electric force in another, or vice-versa. Together, these two fields make up the electromagnetic field, which is best known for underlying light and other electromagnetic waves.

B AND H

There are two quantities that physicists may refer to as the magnetic field, notated H and B. Although the term "magnetic field" was historically reserved for H, with B being termed the "magnetic induction", is now understood to be the more fundamental entity. Modern writers vary in their usage of B as the magnetic field. The convention of referring to B as the magnetic field and the more fundamental B magnetic field, before treating the H field. But the reader is cautioned that the literature is inconsistent. A technical paper may fail to make a distinction between the magnetic field and magnetic induction, knowing that the audience may know the difference.

Alternative names for B and H

The vector field H is known among electrical engineers as the magnetic field intensity or magnetic field strength and is also known among physicists as auxiliary magnetic field or magnetizing field. The vector field B is known among electrical engineers as magnetic flux density or magnetic induction or simply magnetic field, as used by physicists.

The magnetic field B has the SI units of teslas (T), equivalent to webers per square meter (Wb/m^2) or volt seconds per square meter (V s/m^2). In cgs units, B has units of gauss

(G). The vector field H is measured in Amperes/meter (A/m) in SI or oersted (Oe) in cgs units.

PERMANENT MAGNETS AND MAGNETIC POLES

Permanent magnets are objects that produce their own persistent magnetic fields. All permanent magnets have both a north and a south pole. Like poles repel and opposite poles attract. The magnetism in a permanent magnet arises from properties of the atoms (in particular the electrons) that compose it.

Each atom acts like a little individual magnet. If these magnets line up, they combine to create a macroscopic magnetic effect. If allowed to twist freely, a magnet will turn to point in the direction of the magnetic field at its location.

A compass is a small magnet that uses this effect to point in the direction of the local magnetic field. By definition, the direction of the magnetic field at a point is the direction that the north pole of a magnet would want to point. If a compass is placed near the north pole of a magnet then it will point away from that pole—like poles repel.

In other words, the magnetic field points away from a magnet near its north pole. The opposite occurs if we place the compass near a magnet's south pole; the magnetic field points towards the magnet near its south pole. Not all magnetic fields are describable in terms of poles, though. A straight current-carrying wire, for instance, produces a magnetic field that points neither towards nor away from the wire, but encircles it instead.

VISUALIZING THE MAGNETIC FIELD

The strength and direction of the magnetic field due to an object varies from position to position. Mapping out this magnetic field is simple in principle. First, measure the strength and direction of the magnetic field at a large number of points. Then mark each location with an arrow (called a vector) pointing in the direction of the magnetic field with a length proportional to the strength of the magnetic field.

This is a valid and useful way of marking out and

visualizing the magnetic field of an object. It has the unfortunate consequence, though, of cluttering up a graph even when using a small number of points. An alternative method of visualizing the magnetic field is to use "magnetic field lines".

Magnetic B Field Lines

Various physical phenomena have the effect of displaying magnetic field lines. For example, iron filings placed in a magnetic field will line up in such a way as to visually show the orientation of the magnetic field. Another place where magnetic fields are visually displayed is in the polar auroras, in which visible streaks of light line up with the local direction of Earth's magnetic field (due to plasma particle dipole interactions). In these phenomena, lines or curves appear that follow along the direction of the local magnetic field.

These field lines provide us with a way to depict or draw the magnetic field (or any other vector field). Technically, field lines are a set of lines through space whose direction at any point is the direction of the local magnetic field, and whose density is proportional to the magnitude of the local magnetic field. Note that when a magnetic field is depicted with field lines, it is *not* meant to imply that the field is only nonzero along the drawn-in field lines.

Rather, the field is typically smooth and continuous everywhere, and can be estimated at *any* point (whether on a field line or not) by looking at the direction and density of the field lines nearby. The choice of which field lines to draw in such a depiction is arbitrary, apart from the requirement that they be spaced out so that their density approximates the magnitude of the local field. The level of detail at which the magnetic field is depicted can be increased by increasing the number of lines.

Field lines are a useful way to represent any vector field and can often be used to reveal sophisticated properties of that field quite simply. One important property of the magnetic B field that can be verified with field lines is that it always makes complete loops. Magnetic field lines neither start nor end (although they can extend to or from infinity). To date no exception to this rule has been found.

Even when it appears that a magnetic field has an end (such as when it leaves near a north pole or enters near a south pole of a magnet) in reality it does not. In the case of the permanent magnet the field lines complete the loop inside of the magnet traveling from the south to the north pole. Since magnetic field lines always come in loops, magnetic poles always come in N and S pairs. If a magnetic field line enters a magnet somewhere it has to leave the magnet somewhere else; it is not allowed to have an end point. For this reason as well, cutting a magnet in half will result in two separate magnets each with both a north and a south pole.

Field lines are also a good tool for visualizing magnetic forces. When dealing with magnetic fields in ferromagnetic substances like iron, and in plasmas, the magnetic forces can be understood by imagining that the field lines exert a tension, (like a rubber band) along their length, and a pressure perpendicular to their length on neighboring field lines. The 'unlike' poles of magnets attract because they are linked by many field lines, while 'like' poles repel because the field lines between them don't meet, but run parallel, pushing on each other. They can also be represented mathematically in terms of "Euler potentials," though that representation is nonlinear and has other problems.

Earth's Magnetic Field

The north pole of earth is near the top of the diagram, the south pole near the bottom. Notice that the south pole of that magnet is deep in Earth's interior below Earth's North Magnetic Pole. Earth's magnetic field is produced in the outer liquid part of its core due to a dynamo that produce electrical currents there. Because of Earth's magnetic field, a compass placed anywhere on Earth will turn so that the "north pole" of the magnet inside the compass points roughly north, toward Earth's north magnetic pole in northern Canada. This is the traditional definition of the "north pole" of a magnet, although other equivalent definitions are also possible. One confusion that arises from this definition is that if Earth itself is considered as a magnet, the *south* pole of that magnet would be the one nearer the north magnetic pole, and vice-versa. (Opposite poles attract and the north pole of the compass magnet is attracted to the north magnetic pole.) The north magnetic pole is so named not because of the polarity of the field there but because of its geographical location.

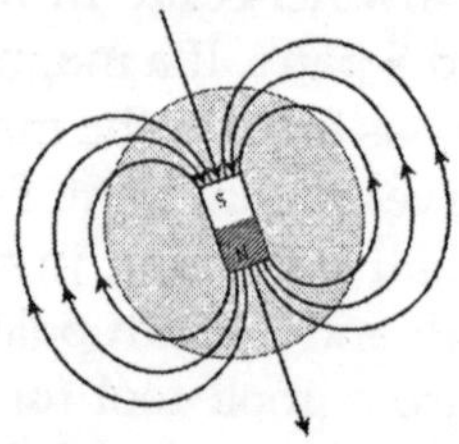

Fig. A Sketch of Earth's Magnetic Field

The right is a sketch of Earth's magnetic field represented by field lines. The magnetic field at any given point does not point straight toward (or away) from the poles and has a significant up/down component for most locations. (In addition, there is an East/West component as Earth's magnetic poles do not coincide exactly with Earth's geological pole.) The magnetic field is as if there were a magnet deep in Earth's interior. Earth's magnetic field is probably due to a dynamo that produces electric currents in the outer liquid part of its core. Earth's magnetic field is not constant: Its strength and the location of its poles vary. The poles even periodically reverse direction, in a process called geomagnetic reversal.

EFFECTS OF THE MAGNETIC FIELD

A magnetic field has many effects on materials and on individual particles. All of these effects can be expressed due to its affects on elementary charges and magnetic dipoles. There are four elementary ways that a magnetic field can affect a charge or a magnetic dipole.

- Sideways force on a moving charge or current.
- Torque on a magnetic dipole.
- Force on a magnetic dipole due to a non-uniform B.

- Force on a charge due to a changing B.

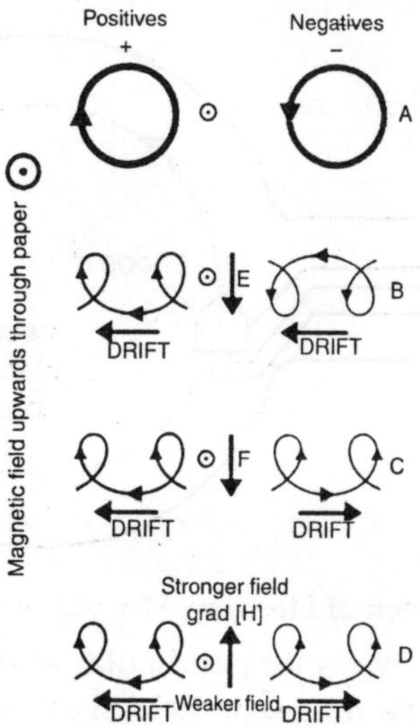

Fig. Charged Particle Drifts in a Homogenous Magnetic Field. (A) No Disturbing Force (B) With an Electric Field, E (C) With an Independent Force, F (e.g. gravity) (D) In an Inhomogeneous Magnetic Field, Grad H

Force Due to a Magnetic Field on a Moving Charge

Force on a Charged Particle

A charged particle moving in a magnetic B field will feel a *sideways* force that is proportional to the strength of the magnetic field, the component of the velocity that is perpendicular to the magnetic field and the charge of the particle. This force is known as the Lorentz force, and is given by;

$$F = q\,(v \times B)$$

where

F is the force (in newtons)

q is the electric charge of the particle (in coulombs)

v is the instantaneous velocity of the particle (in meters per second)

B is the magnetic field (in teslas).

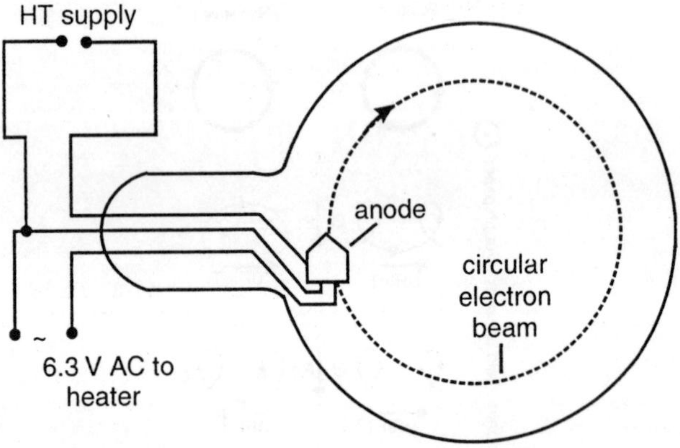

Fig. Beam of Electrons Moving in a Circle

The force is always perpendicular to both the velocity of the particle and the magnetic field that created it. Neither a stationary particle nor one moving in the direction of the magnetic field lines will experience a force. For that reason, charged particles move in a circle around magnetic field lines; this is called cyclotron motion.

Because the magnetic field is always perpendicular to the motion, the magnetic fields can do no work on a charged particle; a magnetic field alone cannot speed up or slow down a charged particle. It can and does, however, change the particle's direction, even to the extent that a force applied in one direction can cause the particle to drift in a perpendicular direction.

Force on Current-carrying Wire

The force on a current carrying wire is similar to that of a moving charge as expected since a charge carrying wire is a collection of moving charges. A current carrying wire will feel a sideways force in the presence of a magnetic field. The Lorentz force on a macroscopic current is often referred to as the *Laplace force.*

The Right-hand Rule: For a conventional current or moving positive charge in the direction of the thumb of the right hand and the magnetic field along the direction of the fingers (pointing away from palm) the force on the current will be in a direction out of the palm. The direction of the force is reversed for a negative charge.

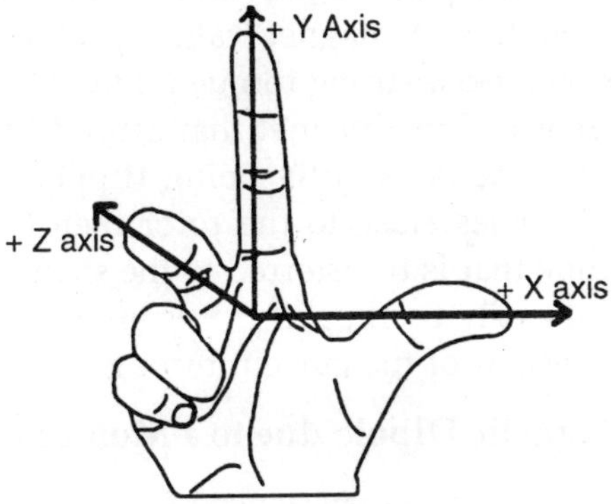

Fig. The Right-hand Rule

Direction of Force

The direction of force on a positive charge or a current is determined by the right-hand rule. Using the right hand and pointing the thumb in the direction of the moving positive charge or positive current and the fingers in the direction of the magnetic field the resulting force on the charge will point outwards from the palm. The force on a negative charged particle is in the opposite direction. If both the speed and the charge are reversed then the direction of the force remains the same.

For that reason a magnetic field measurement cannot distinguish whether there is a positive charge moving to the right or a negative charge moving to the left. On the other hand, a magnetic field combined with an electric field *can* distinguish between these an alternative, similar trick to the right hand rule is Fleming's left hand rule.

Torque on a Magnetic Dipole

A magnet placed in a magnetic field will feel a torque that will try to align the magnet with the magnetic field. The torque on a magnet due to an external magnetic field is easy to observe by placing two magnets near each other while allowing one to rotate. This magnetic torque is the basis for how compasses work. It is used to define the direction of the magnetic field.

The magnetic torque also provides the driving torque for simple electric motors. A magnet (called a rotor) placed on a rotating shaft will feel a strong torque if like poles are placed near its own poles. If the magnet that caused the rotation—called the stator—is constantly being flipped such that it always has like poles close to the rotor then the rotor will generate a torque that is transferred to the shaft. The polarity of the rotor can easily be flipped if it is an electromagnet by flipping the direction of the current through its coils.

Force on a Magnetic Dipole due to a Mon-uniform B

The most commonly experienced effect of the magnetic field is the force between two magnets: Like poles repel and opposites attract. One can, in fact, express this force in terms of the pole locations and strengths (pole distributions) in the two magnets attracting and repulsing each other. This model is called the "Gilbert model" and produces both the correct force between two magnets, and the correct B field outside of the magnets, but the wrong magnetic field *inside* the magnets.

Physically accurate picture would be based on the fundamental fact that a magnetic dipole experiences a force, when placed in a *non-uniform* external magnetic field. (In a uniform field, it will experience a torque but no force.) The south pole of one magnet is attracted to the north pole of another magnet because of the specific way in which each of the microscopic dipoles in either magnet responds to the non-uniform field of the other magnet.

The force on a magnetic dipole does not depend directly on the strength or direction of the magnetic field, but only on how these vary with location. A magnet will move to maximize the magnetic field in the direction of its magnetic moment.

Care should be taken to distinguish the magnetic force on a magnetic dipole from the magnetic force on a moving charge. The magnetic force on a charge only occurs when the charge is moving and is in a sideways direction. It is felt for both uniform and non-uniform magnetic fields. The magnetic force on a dipole, on the other hand, is present only in non-uniform (in space) fields and is in the direction that increases the component of the magnetic field in the direction parallel to the dipole's magnetic moment. Neither does the force on a magnetic dipole depend on its speed.

Electric Force due to a Changing B

If the magnetic field in an area is varying with time it generates an electric field that forms closed loops around that area. A conducting wire that forms a closed loop around the area will have an induced voltage generated by this changing magnetic field.

This effect is represented mathematically as Faraday's Law and forms the basis of many generators. Care must be taken to understand that the changing magnetic field is a source for an *extended* electric field. The changing magnetic field does not only create an electric field at that location; rather it generates an electric field that forms closed loops around the location where the magnetic field is changing.

Mathematically, Faraday's law is most often represented in terms of the change of magnetic flux with time. The magnetic flux is the property of a closed loop (say of a coil of wire) and is the product of the area times the magnetic field that is normal to that area. Engineers and physicists often use magnetic flux as a convenient physical property of a loops.

They then express the magnetic B field as the magnetic flux per unit area. It is for this reason that the B field is often referred to as the "magnetic flux density". This approach has the benefit of making certain calculations easier such as in magnetic circuits. It is typically not used outside of electrical circuits, though, because the magnetic B field truly is the more 'fundamental' quantity in that it directly connects all of electrodynamics in the simplest manner.

SOURCES OF MAGNETIC FIELDS

Magnetic fields can be created in a number of different

ways. All of these ways are based on three elementary ways to create a magnetic B field.

- Electrical currents (moving charges)
- Magnetic dipoles
- Changing electric field

These sources are thought to affect the virtual particles that compose the field.

Electrical Currents

All moving charges produce a magnetic field. The magnetic field of a moving charge is very complicated but is well known. It forms closed loops around a line that is pointing in the direction the charge is moving. The magnetic field of a current on the other hand is much easier to calculate.

Magnetic Field of a Steady Current

Current (*I*) through a wire produces a magnetic field (B) around the wire. The field is oriented according to the right hand grip rule.

The magnetic field generated by a *steady current* (a continual flow of charges, for example through a wire, which is constant in time and in which charge is neither building up nor depleting at any point), is described by the Biot-Savart law. This is a consequence of Ampere's law, one of the four Maxwell's equations that describe electricity and magnetism. The magnetic field lines generated by a current carrying wire form concentric circles around the wire. The direction of the magnetic field of the loops is determined by the right hand grip rule. The strength of the magnetic field decreases with distance from the wire.

A current carrying wire can be bent in a loop such that the field is concentrated (and in the same direction) inside of the loop. The field will be weaker outside of the loop. Stacking many such loops to form a solenoid (or long coil) can greatly increase the magnetic field in the centre and decrease the magnetic field outside of the solenoid.

Such devices are called electromagnets and are extremely important in generating strong and well controlled magnetic fields. An infinitely long solenoid will have a uniform magnetic field inside of the loops and no magnetic field outside. A finite length electromagnet will produce essentially the same magnetic field as a uniform permanent magnet of the same shape and size. An electromagnet has the advantage, though, that you can easily vary the strength (even creating a field in the opposite direction) simply by controlling the input current. One important use is to continually switch the polarity of a stationary electromagnet to force a rotating permanent magnet to continually rotate using the fact that opposite poles attract and like poles repel. This can be used to create an important type of electrical motor.

Magnetic Dipoles

The magnetic field due to a permanent magnet is well known. But, what causes the magnetic field of a permanent magnet? The answer again is that the magnetic field is essentially created due to currents. But this time it is due to the cumulative effect of many small 'currents' of electrons 'orbiting' the nuclei of the magnetic material. Alternatively it is due to the structure of the electron itself which, in some sense, can be thought of as forming a tiny loop of current. (The true nature of the electron's magnetic field is relativistic in nature, but this model often works.)

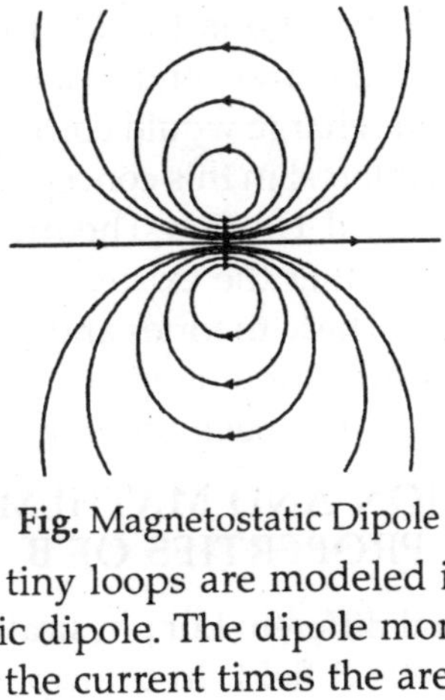

Fig. Magnetostatic Dipole

Both of these tiny loops are modeled in terms of what is called the magnetic dipole. The dipole moment of that dipole can be defined as the current times the area of the loop, then an equation for the magnetic field due to that magnetic dipole can be derived. Magnetic field of a larger magnet can be calculated by adding up the magnetic fields of many magnetic dipoles.

Changing Electric Field

The final known source of magnetic fields is a changing electric field. Just as a changing magnetic field generates an electric field so does a changing electric field generate a magnetic field. (These two effects bootstrap together to form electromagnetic waves, such as light.) Similar to the way magnetic field lines form close loops around a current a time varying electric field generates a magnetic field that forms closed loops around the region where the electric field is changing. The strength of this magnetic field is proportional to the time rate of the change of the electric field (which is called the displacement current). The fact that a changing electric field creates a magnetic field is known as Maxwell's correction to Ampere's Law.

Magnetic Monopole

The magnetic monopole is a hypothetical particle (it may or may not exist). A magnetic monopole would have, as its name suggests, only one pole. In other words, it would possess "magnetic charge" analogous to electric charge. Positive magnetic charge would correspond to an isolated north pole, and negative magnetic charge would correspond to an isolated south pole. Modern interest in this concept stems from particle theories, notably Grand Unified Theories and superstring theories, that predict either the existence or the possibility of magnetic monopoles. These theories and others have inspired extensive efforts to search for monopoles. Despite these efforts, no magnetic monopole has been observed to date.

DEFINITION AND MATHEMATICAL PROPERTIES OF B

There are several different but physically equivalent ways to define the magnetic B field. In principle any of the above effects due to the magnetic field or any of the sources of the magnetic field can be used to define its magnitude and the direction. Its direction at a given point can be thought of as being the direction that a *hypothetical* freely rotating small test dipole would rotate to point if it *were* placed at that point. Its magnitude is defined (in SI units) in terms of the voltage induced per unit area on a current carrying loop in a uniform magnetic field normal to the loop when the magnetic field is reduced to zero in a unit amount of time. The SI unit of magnetic field is the Tesla. The magnetic field vector is a pseudovector (also called an axial vector). (This is a technical statement about how the magnetic field behaves when you reflect the world in a mirror.) This fact is apparent from many of the definitions and properties of the field; for example, the magnitude of the field is proportional to the torque on a dipole, and torque is a well-known pseudovector.

Maxwell's Equations

The magnetic field is a vector field. (The magnetic field at each point in space and time is represented by its own vector.) As a vector field, the magnetic field has two important mathematical properties. These properties, along with the corresponding properties of the electric field, make up Maxwell's Equations.

The first is that the magnetic B field never starts nor ends at a point. Whatever magnetic field lines enter a region has to eventually leave that region. This is mathematically equivalent to saying that the divergence of the magnetic B is zero. (Such vector fields are called solenoidal vector fields.) This property is called Gauss' law for magnetism and is one of Maxwell's Equations. It is also equivalent to the statement that there are no magnetic monopoles.

The second mathematical property of the magnetic field is that it always loops around the source that creates it. This source could be a current, a magnet, or a changing electric field, but it is always within the loops of magnetic field they create. Mathematically, this fact is described by the Ampère-Maxwell equation.

MEASURING THE MAGNETIC B FIELD

There are many ways of measuring the magnetic B field, many of which use the effects described above. Devices used to measure the local magnetic field are called magnetometers. Important magnetometers include using a rotating coil, Hall effect magnetometers, NMR magnetometer, SQUID magnetometer, and a fluxgate magnetometer. The magnetic fields of distant astronomical objects can be determined by noting their effects on local charged particles. For instance, electrons spiraling around a field line will produce synchotron radiation which is detectable in radio waves.

Hall Effect

Because the Lorentz force is charge-sign-dependent it results in charge separation when a conductor with current is placed in a transverse magnetic field, with a buildup of opposite charges on two opposite sides of conductor in the direction normal to the magnetic field, and the potential difference between these sides can be measured. The Hall effect is often used to measure the magnitude of a magnetic field as well as to find the sign of the dominant charge carriers in semiconductors (negative electrons or positive holes).

Squid Magnetometer

Superconductors are materials with both distinctive electric properties (perfect conductivity) and magnetic properties (such as the Meissner effect, in which many superconductors can perfectly expel magnetic fields). Due to these properties, it turns out that loops that incorporate superconducting material and their Josephson junctions can function as very sensitive magnetometers, called SQUIDs.

THE H FIELD

The term 'magnetic field' can also be used to describe the magnetic H field. The magnetic H field is similar to B in that it is a vector field, but its units are often different. In SI units, B and H are measured in teslas (T) and amperes per meter (A/m), respectively; or, in cgs units, in gauss (G) and oersteds (Oe), respectively. Outside of magnetizable materials, the two fields are identical (apart from possibly a constant conversion factor), but inside a magnetic material they can differ substantially. Quantitatively, the fields are related by the following equations:

$$B = \mu_0 (H + M) \text{ (SI units)}$$

$$B = H + 4\pi M \text{ (cgs units),}$$

Where is M magnetization density of any magnetic material.

Physical Interpretation of the H Field

When magnetic materials are present, the total magnetic field B is caused by two different types of currents which need to be distinguished: free current and bound current. Free currents are the ordinary currents in wires and other conductors, that can be controlled and measured. Bound currents are the tiny circular currents inside atoms that are responsible for the magnetization M of magnetic materials.

Although *t* he actual source of the magnetic field in electron orbitals of atoms is complex, the magnetic properties of a material can be accounted for by assuming it is divided into tiny blocks, each of which has a current flowing around its outside surface, perpendicular to the magnetic field axis. As an example of bound current consider a uniform permanent bar magnet.

A piece of iron is formed of many tiny regions called magnetic domains, each of which is a magnetic dipole, essentially a tiny loop of current. In a bar magnet, most of these dipoles have their poles lined up, creating a large magnetic field. If we add up the currents of all these tiny loops we will find that the currents cancel in the interior of the material, but add up along the sides of the bar. This current loops around the sides and not at the poles. No one charge makes the complete trip around the magnet each charge is bound to its tiny loop but the net effect is exactly equivalent to a real current that flows around the outside surface of the magnet, perpendicular to the magnetic field axis. If the magnetization is not uniform then a bound current will flow through the bulk of the magnetic material as well.

The magnetic H is useful because it treats these two types of currents differently. The free currents it treats in the normal fashion and therefore has the same form as the magnetic B field it would generate. The magnetic H fields treats the field inside of a magnetic material (due to that magnetic material) in a manner similar to the Gilbert model. By subtracting the magnetization from the B field we are essentially converting the bound current sources to Gilbert-like magnetic charges at the poles. Unlike the magnetic B, which always forms closed loops, the field due to the magnetic charges flow outward (or inward depending on the sign of the magnetic charge) in both directions from the poles. And while the magnetic field is exactly the same on the outside of the magnetic material for both models the magnetic fields inside are quite different.

Putting both sources together we see that the magnetic H field is the same as the magnetic B field to a multiplicative constant outside of magnetic materials, but is completely different from the magnetic B field inside a magnetic material. The advantage of this hybrid H field is that these sources are treated so differently that we can often pick out one source from the other. For example a line integral of the magnetic H field in a closed loop will yield the total free current in the loop (and not the bound current). This is unlike the magnetic B field where a similar integral will yield the sum of both the free and the bound current. If one wants to isolate the contribution due to the bound currents then a surface integral of H over any closed surface will pick out the 'magnetic charges' at the poles.

Sources of the H Field

Unlike the magnetic B field that only has a current source such that the magnetic B field loops around currents, the magnetic H field has two types of sources. The first source of magnetic H field are the *free* currents for which H loop around similar to the way B field loops around the total current. The second source of the magnetic H field are 'magnetic charges' near the poles of the magnetic material. More precisely, these 'magnetic charges' are calculated as $-\nabla.M$.

Uses of the H Field

Energy Stored in Magnetic Fields

In order to create a magnetic field we need to do work to establish a free current. If we are to ask how much energy does it take to create a specific magnetic field using a particular free current then we need to distinguish between the free and the bound currents. It is the free current that we are 'pushing' on. The bound currents are freeloaders.

They create a magnetic field that the free current has to work against without doing any of the work. If we are to calculate the energy of creating a magnetic field B we need to have a way of separating out the free current. The magnetic B cannot be used to determine this free current since B does not distinguish between bound and free current.

The magnetic H field does treat the two sources differently. Therefore it is useful in calculating the energy needed to create a magnetic field with a free current in the presence of magnetic materials. In this case the energy density needed, assuming a linear relationship between H and B, has the form of:

$$u = \frac{H.B}{2}.$$

If there are no magnetic materials around then we can replace H with $\frac{B}{\mu_0}$,

$$u = \frac{B.B}{2\mu_0}.$$

Magnetic Circuits

A second use for H is in magnetic circuits where inside a linear material B = μH. Here, μ is the permeability of the material. This is similar in form to Ohm's Law J = σE, where J is the current density, σ is the conductance and E is the Electric field. Extending this analogy we derive the counterpoint to the macroscopic Ohm's law $\left(\frac{V}{R} = I\right)$ as:

$$\Phi = \frac{F}{R_m},$$

Where $\Phi = \int B.dA$ is the magnetic flux in the circuit, $F = \int H.dl$ is the magnetomotive force applied to the circuit, and R_m is the reluctance of the circuit. Here the reluctance R_m is a quantity similar in nature to resistance for the flux.

Using this analogy it is straight-forward to calculate the magnetic flux of complicated magnetic field geometries, by using all the available techniques of circuit theory. The difference between the B and the H vectors can be traced back to Maxwell's 1855 paper entitled *on Faraday's Lines of Force.* It is later clarified in his concept of a sea of molecular vortices that appears in his 1861 paper on Physical Lines of Force-1861.

Within that context, H represented pure vorticity (spin), whereas B was a weighted vorticity that was weighted for the density of the vortex sea. Maxwell considered magnetic permeability μ to be a measure of the density of the vortex sea. Hence the relationship,

- Magnetic induction current causes a magnetic current density

$$B = \mu H$$

was essentially a rotational analogy to the linear electric current relationship,

- Electric convection Current

$$j = \rho v$$

where ñ is electric charge density. B was seen as a kind of magnetic current of vortices aligned in their axial planes, with H being the circumferential velocity of the vortices. With μ representing vortex density, we can now see how the product of μ with vorticity H leads to the term magnetic flux density which we denote as B.

The electric current equation can be viewed as a convective current of electric charge that involves linear motion. By analogy, the magnetic equation is an inductive current involving spin. There is no linear motion in the inductive current along the direction of the B vector. The magnetic inductive current represents lines of force. In particular, it represents lines of inverse square law force. The extension of the above considerations confirms that where B is to H, and where J is to ρ, then it necessarily follows from Gauss's law and from the equation of continuity of charge that E is to D. i.e., B parallels with E, whereas H parallels with D.

ROTATING MAGNETIC FIELDS

The rotating magnetic field is a key principle in the operation of alternating-current motors. A permanent magnet in such a field will rotate so as to maintain its alignment with the external field. This effect was conceptualized by Nikola Tesla, and later utilised in his, and others', early AC (alternating-current) electric motors. A rotating magnetic field can be constructed using two orthogonal coils with 90 degrees phase difference in their AC currents. However, in practice such a system would be supplied through a three-wire arrangement with unequal currents.

This inequality would cause serious problems in standardization of the conductor size and so, in order to overcome it, three-phase systems are used where the three currents are equal in magnitude and have 120 degrees phase difference. Three similar coils having mutual geometrical angles of 120 degrees will create the rotating magnetic field in this case. The ability of the three-phase system to create a rotating field, utilized in electric motors, is one of the main reasons why three-phase systems dominate the world's electrical power supply systems.

Because magnets degrade with time, synchronous motors and induction motors use short-circuited rotors (instead of a magnet) following the rotating magnetic field of a multicoiled stator. The short-circuited turns of the rotor develop eddy currents in the rotating field of the stator, and these currents in turn move the rotor by the Lorentz force.

In 1882, Nikola Tesla identified the concept of the rotating magnetic field. In 1885, Galileo Ferraris independently researched the concept. In 1888, Tesla gained U.S. Patent 381,968 for his work. Also in 1888, Ferraris published his research in a paper to the *Royal Academy of Sciences* in Turin.

SPECIAL RELATIVITY AND ELECTROMAGNETISM

Moving Magnet and Conductor Problem

Imagine a moving conducting loop that is passing by a stationary magnet. Such a conducting loop will have a current generated in it as it passes through the magnetic field. But why? It is answering this seemingly innocent question that led Albert Einstein to develop his theory of special relativity.

A stationary observer would see an unchanging magnetic field and a moving conducting loop. Since the loop is moving all of the charges that make up the loop are also moving. Each of these charges will have a sideways, Lorentz force, acting on it which generates the current. Meanwhile, an observer on the moving reference frame would see a *changing* magnetic field and *stationary* charges. (The loop is not moving in this observers reference frame. This changing magnetic field generates an *electric* field.

The stationary observer claims there is *only* a magnetic field that creates a *magnetic force* on a moving charge. The moving observer claims that there is both a magnetic and an electric field but all of the force is due to the *electric* field. Which is true? Does the electric field exist or not? The answer, according to special relativity, is that both observers are right from their reference frame. A pure magnetic field in one reference can be a mixture of magnetic and electric field in another reference frame.

Electrical Fields Different Aspects of Same Phenomenon

According to special relativity, electric and magnetic forces are part of a single physical phenomenon, electromagnetism; an electric force perceived by one observer will be perceived by another observer in a different frame of reference as a mixture of electric and magnetic forces. A magnetic force can be considered as simply the relativistic part of an electric force when the latter is seen by a moving observer.

More specifically, rather than treating the electric and magnetic fields as separate fields, special relativity shows that they naturally mix together into a rank-2 tensor, called the electromagnetic tensor. This is analogous to the way that special relativity "mixes" space and time into spacetime, and mass, momentum and energy into four-momentum.

MAGNETOHYDRODYNAMICS

Magnetohydrodynamics (MHD) (*magnetofluiddynamics* or *hydromagnetics*) is the academic discipline which studies the dynamics of electrically conducting fluids. Examples of such fluids include plasmas, liquid metals, and salt water. The word *magnetohydrodynamics (MHD)* is derived from *magneto-* meaning magnetic field, and *hydro-* meaning liquid, and *-dynamics* meaning movement. The field of MHD was initiated by Hannes Alfvén, for which he received the Nobel Prize in Physics in 1970.

The idea of MHD is that magnetic fields can induce currents in a moving conductive fluid, which create forces on the fluid, and also change the magnetic field itself. The set of equations which describe MHD are a combination of the Navier-Stokes equations of fluid dynamics and Maxwell's equations of electromagnetism. These differential equations have to be solved simultaneously, either analytically or numerically. MHD is a *fluid* theory and as such it cannot treat kinetic phenomena, i.e. those in which the existence of discrete particles or of a non-thermal velocities distribution, are important.

Ideal and Resistive MHD

The simplest form of MHD, Ideal MHD, assumes that the fluid has so little resistivity that it can be treated as a perfect conductor. This is the limit of infinite magnetic Reynolds number. In ideal MHD, Lenz's law dictates that the fluid is in a sense *tied* to the magnetic field lines. To explain, in ideal MHD a small rope-like volume of fluid surrounding a field line will continue to lie along a magnetic field line, even as it is twisted and distorted by fluid flows in the system.

The connection between magnetic field lines and fluid in ideal MHD fixes the topology of the magnetic field in the fluid—for example, if a set of magnetic field lines are tied into a knot, then they will remain so as long as the fluid/plasma has negligible resistivity.

This difficulty in reconnecting magnetic field lines makes it possible to store energy by moving the fluid or the source of the magnetic field. The energy can then become available if the conditions for ideal MHD break down, allowing magnetic reconnection that releases the stored energy from the magnetic field.

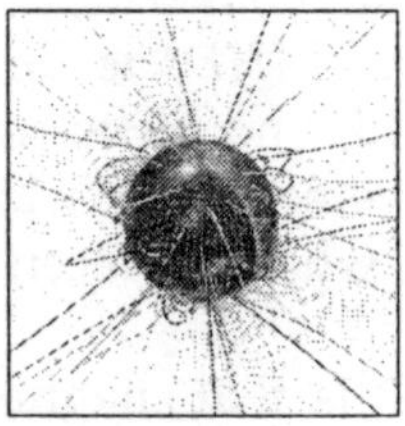

Fig. Resistive MHD

Ideal MHD Equations

The ideal MHD equations consist of the continuity equation, the momentum equation, and Ampere's Law in the limit of no electric field and no electron diffusivity, and a temperature evolution equation. As with any fluid description to a kinetic system, a closure approximation must be applied to highest moment of the particle distribution equation. This is often accomplished with approximations to the heat flux through a condition of adiabaticity or isothermality.

Applicability of Ideal MHD to Plasmas

Ideal MHD is only strictly applicable when:

- The plasma is strongly collisional, so that the time scale of collisions is shorter than the other characteristic times in the system, and the particle distributions are therefore close to Maxwellian.
- The resistivity due to these collisions is small. In particular, the typical magnetic diffusion times over any scale length present in the system must be longer than any time scale of interest.

- We are interested in length scales much longer than the ion skin depth and Larmor radius perpendicular to the field, long enough along the field to ignore Landau damping, and time scales much longer than the ion gyration time (system is smooth and slowly evolving).

The Importance of Resistivity

In an imperfectly conducting fluid the magnetic field can generally move through the fluid following a diffusion law with the resistivity of the plasma serving as a diffusion constant. This means that solutions to the ideal MHD equations are only applicable for a limited time for a region of a given size before diffusion becomes too important to ignore. One can estimate the diffusion time across a solar active region (from collisional resistivity) to be hundreds to thousands of years, much longer than the actual lifetime of a sunspot—so it would seem reasonable to ignore the resistivity. By contrast, a meter-sized volume of seawater has a magnetic diffusion time measured in milliseconds.

Even in physical systems which are large and conductive enough that simple estimates suggest that we can ignore the resistivity, resistivity may still be important: many instabilities exist that can increase the effective resistivity of the plasma by factors of more than a billion. The enhanced resistivity is usually the result of the formation of small scale structure like current sheets or fine scale magnetic turbulence, introducing small spatial scales into the system over which ideal MHD is broken and magnetic diffusion can occur quickly.

When this happens, magnetic reconnection may occur in the plasma to release stored magnetic energy as waves, bulk mechanical acceleration of material, particle acceleration, and heat. Magnetic reconnection in highly conductive systems is important because it concentrates energy in time and space, so that gentle forces applied to a plasma for long periods of time can cause violent explosions and bursts of radiation.

When the fluid cannot be considered as completely conductive, but the other conditions for ideal MHD are satisfied, it is possible to use an extended model called resistive MHD. This includes an extra term in Ampere's Law which models the collisional resistivity. Generally MHD computer simulations are at least somewhat resistive because their computational grid introduces a numerical resistivity.

The Importance of Kinetic Effects

Another limitation of MHD (and fluid theories in general) is that they depend on the assumption that the plasma is strongly collisional so that the time scale of collisions is shorter than the other characteristic times in the system, and the particle distributions are Maxwellian. This is usually not the case in fusion, space and astrophysical plasmas. When this is not the case, or we are interested in smaller spatial scales, it may be necessary to use a kinetic model which properly accounts for the non-Maxwellian shape of the distribution function. However, because MHD is very simple and captures many of the important properties of plasma dynamics it is often qualitatively accurate and is almost invariably the first model tried. Effects which are essentially kinetic and not captured by fluid models include double layers, Landau damping, a wide range of instabilities, chemical separation in space plasmas and electron runaway.

EXTENSIONS TO MAGNETOHYDRODYNAMICS

Resistive MHD

Resistive MHD describes magnetized fluids with finite electron diffusivity ($\eta \neq 0$). This diffusivity leads to a breaking in the magnetic topology.

Extended MHD

Extended MHD describes a class of phenomena in plasmas that are higher order than resistive MHD, but which can adequately be treated with a single fluid description. These include the effects of Hall physics, electron pressure gradients, finite Larmor Radii in the particle gyromotion, and electron inertia.

Two-Fluid MHD

Two-Fluid MHD describes plasmas that include a non-negligible electric field. As a result, the electron and ion momenta must be treated separately. This description is more closely tied to Maxwell's equations as an evolution equation for the electric field exists.

Hall MHD

In 1960, M. J. Lighthill criticized the applicability of ideal or resistive MHD theory for plasmas. It concerned the neglect of the "Hall current term", a frequent simplification made in magnetic fusion theory. Hall-magnetohydrodynamics (HMHD) takes into account this electric field description of magnetohydrodynamics. The most important difference is that in the absence of field line breaking, the magnetic field is tied to the electrons and not to the bulk fluid.

Collisionless MHD

MHD is also often used for collisionless plasmas. In that case the MHD equations are derived from the Vlasov equation.

APPLICATIONS OF MAGNETOHYDRODYNAMICS

Geophysics

The fluid core of the Earth and other planets is theorized to be a huge MHD dynamo that generates the Earth's magnetic field due to the motion of liquid iron.

Astrophysics

MHD applies quite well to astrophysics since over 99% of baryonic matter content of the Universe is made up of plasma, including stars, the interplanetary medium (space between the planets), the interstellar medium (space between the stars), nebulae and jets. Many astrophysical systems are not in local thermal equilibrium, and therefore require an additional kinematic treatment to describe all the phenomena within the system.

Sunspots are caused by the Sun's magnetic fields, as

Joseph Larmor theorized in 1919. The solar wind is also governed by MHD. The differential solar rotation may be the long term effect of magnetic drag at the poles of the Sun, an MHD phenomenon due to the Parker spiral shape assumed by the extended magnetic field of the Sun.

Previously, theories describing the formation of the Sun and planets could not explain how the Sun has 99.87% of the mass, yet only 0.54% of the angular momentum in the solar system. In a closed system such as the cloud of gas and dust from which the Sun was formed, mass and angular momentum are both conserved.

That conservation would imply that as the mass concentrated in the centre of the cloud to form the Sun, it would spin up, much like a skater pulling their arms in. The high speed of rotation predicted by early theories would have flung the proto-Sun apart before it could have formed. However, magnetohydrodynamic effects transfer the Sun's angular momentum into the outer solar system, slowing its rotation. Breakdown of ideal MHD (in the form of magnetic reconnection) is known to be the cause of solar flares, the largest explosions in the solar system. The magnetic field in a solar active region over a sunspot can become quite stressed over time, storing energy that is released suddenly as a burst of motion, X-rays, and radiation when the main current sheet collapses, reconnecting the field.

AMPERE'S CIRCUITAL LAW

"Ampere's law" redirects here. For the law describing forces between current-carrying wires. In classical electromagnetism, Ampere's circuital law, discovered by Andre-Marie Ampere in 1826, relates the integrated magnetic field around a closed loop to the electric current passing through the loop. It is one of Maxwell's equations, which form the basis of classical electromagnetism.

Original Ampere's Circuital law

In its historically original form, Ampere's Circuital Law relates the magnetic field to its electric current source. The law can be written in two forms, the "integral form" and the "differential form". The forms are equivalent, and related by the Kelvin-Stokes theorem.

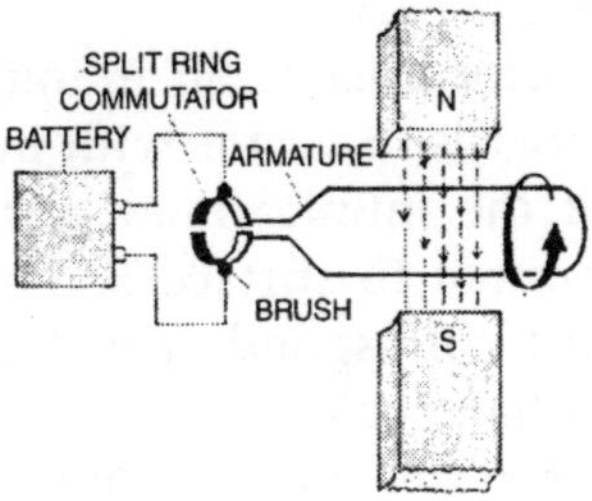

Fig. An Electric Current Produces a Magnetic Field

Integral Form

In SI units, the "integral form" of the original Ampere's circuital law is:

$$\oint_C B.d\ell = \mu_0 \iint_S J_f.dS$$

or equivalently,

$$\oint_C B.d\ell = \mu_0 I_{enc}$$

where:

- $\oint_C$ is the closed line integral around the closed curve C.
- B is the magnetic field in tesla.
- is the vector dot product.
- $d\ell$ is an infinitesimal element (differential) of the curve *C* (i.e. a vector with magnitude equal to the length of the infinitesimal line element, and direction given by the tangent to the curve *C*,
- $\iint_S$ denotes an integral over the surface *S* enclosed by the curve *C*. The double integral sign is meant simply to denote that the integral is two-dimensional in nature.

- μ_0 is the magnetic constant also called the absolute permeability of free space.
- J_f is the free current density through the surface S enclosed by the curve C.
- dS is the vector area of an infinitesimal element of surface S (that is, a vector with magnitude equal to the area of the infinitesimal surface element, and direction normal to surface S. The direction of the normal must correspond with the orientation of C by the right hand rule),
- I_{enci} is the net free current that penetrates through the surface S.

There are a number of ambiguities in the above definitions that warrant elaboration.

First, three of these terms are associated with sign ambiguities: the line integral $\oint_C$ could go around the loop in either direction (clockwise or counterclockwise); the vector area dS could point in either of the two directions normal to the surface; and I_{enc} is the net current passing through the surface S, meaning the current passing through in one direction, minus the current in the other direction—but either direction could be chosen as positive. These ambiguities are resolved by the right-hand rule: When the index-finger of the right-hand points along the direction of line-integration, the outstretched thumb points in the direction that must be chosen for the vector area dS, and current passing in that same direction must be counted as positive. The right hand grip rule can also be used to determine the signs.

Second, there are infinitely many possible surfaces S that have the curve C as their border. Which of those surfaces is to be chosen? If the loop does not lie in a single plane, for example, there is no one obvious choice. The answer is that it does not matter; it can be proven that any surface with boundary C can be chosen.

Differential Form

By the Kelvin-Stokes theorem, this equation can also be

written in a "differential form". Again, this equation only applies in the case where the electric field is constant in time. In SI units, the equation states:

$$\nabla \times B = \mu_0 J_f$$

where

$\nabla \times$ is the curl operator.

FREE CURRENT VERSUS BOUND CURRENT

In contrast, "bound current" arises in the context of bulk materials that can be magnetized and/or polarized. When a material is magnetized (for example, by placing it in an external magnetic field), the electrons remain bound to their respective atoms, but behave as if they were orbiting the nucleus in a particular direction, creating a microscopic current. When the currents from all these atoms are put together, they create the same effect as a macroscopic current, circulating perpetually around the magnetized object. This magnetization current J_M is one contribution to "bound current".

The other source of bound current is bound charge. When an electric field is applied, the positive and negative bound charges can separate over atomic distances in polarizable materials, and when the bound charges move, the polarization changes, creating another contribution to the "bound current", the polarization current J_P.

The total current density J due to free and bound charges is then:

$$J = J_f + J_M + J_P,$$

with J_f the "free" or "conduction" current density.

All current is fundamentally the same, microscopically. Nevertheless, there are often practical reasons for wanting to treat bound current differently from free current. For example, the bound current usually originates over atomic dimensions, and one may wish to take advantage of a simpler theory intended for larger dimensions. The result is that the more microscopic Ampere's law, expressed in terms of B and the microscopic current (which includes free, magnetization and polarization currents), is sometimes put into the equivalent form below in terms of H and the free current only. For a detailed definition of free current and bound current, and the proof that the two formulations are equivalent.

AMPERE'S CIRCUITAL LAW

There are two important issues regarding Ampere's law that require closer scrutiny. First, there is an issue regarding the continuity equation for electrical charge. There is a theorem in vector calculus that states the divergence of a curl must always be zero. Hence $\nabla \cdot \nabla (\times B) = 0$ and so the original Ampere's law implies that $\nabla \cdot J = 0$. But in general $\nabla \cdot J = \partial r / \partial t$, which is non-zero for a time-varying charge density. An example occurs in a capacitor circuit where time-varying charge densities exist on the plates. Second, there is an issue regarding the propagation of electromagnetic waves. For example, in free space, where $J = 0$, Ampere's law implies that $\nabla \times B = 0$, but instead $\nabla \times B = -(1/c^2)\ \partial E/\partial t$.

To treat these situations, the contribution of displacement current must be added to the "free current" term in Ampere's law. James Clerk Maxwell conceived of displacement current as a polarization current in the dielectric vortex sea, which he used to model the magnetic field hydrodynamically and mechanically. He added this displacement current to Ampere's circuital law at equation in his 1861 paper on Physical Lines of Force.

Displacement Current

In free space, the displacement current is related to the time rate of change of electric field. In a dielectric the above contribution to displacement current is present too, but a major contribution to the displacement current is related to the polarization of the individual molecules of the dielectric material. Even though charges cannot flow freely in a dielectric, the charges in molecules can move a little under the influence of an electric field. The positive and negative charges in molecules separate under the applied field, causing an increase in the state of polarization, expressed as the polarization density P. A changing state of polarization is equivalent to a current.

Both contributions to the displacement current are combined by defining the displacement current as:

$$J_D = \frac{\partial}{\partial t} D(r,t),$$

where the electric displacement field is defined as:

$$D = \varepsilon_0 E + P,$$

where ε_0 is the electric constant and P is the polarization density. Substituting this form for D in the expression for displacement current, it has two components:

$$J_D = \varepsilon_0 \frac{\partial E}{\partial t} + \frac{\partial P}{\partial t}.$$

The first term on the right hand side is present everywhere, even in a vacuum. It doesn't involve any actual movement of charge, but it nevertheless has an associated magnetic field, as if it were an actual current. Some authors apply the name *displacement current* to only this contribution.

The second term on the right hand side is the displacement current as originally conceived by Maxwell, associated with the polarization of the individual molecules of the dielectric material. Maxwell's original explanation for displacement current focused upon the situation that occurs in dielectric media. In the modern post-aether era, the concept has been extended to apply to situations with no material media present, for example, to the vacuum between the plates of a charging vacuum capacitor. The displacement current is justified today because it serves several requirements of an electromagnetic theory: correct prediction of magnetic fields in regions where no free current flows; prediction of wave propagation of electromagnetic fields; and conservation of electric charge in cases where charge density is time-varying.

Maxwell-Ampère Equation

Next Ampere's equation is extended by including the polarization current, thereby remedying the limited applicability of the original Ampere's circuital law.

Treating free charges separately from bound charges, Ampere's equation including Maxwell's correction in terms of the H-field is (the H-field is used because it includes the magnetization currents, so J_M does not appear explicitly:

$$\oint_C H.d\ell = \iint_S \left(J_f + \frac{\partial}{\partial t} D \right).dA$$

(integral form), where:

- H is the magnetic H field (also called "auxiliary magnetic field", "magnetic field intensity", or just "magnetic field",
- D is the electric displacement field
- J_f is the enclosed conduction current or free current density,

or (differential form):

$$\nabla \times H = J_f + \frac{\partial}{\partial t} D.$$

On the other hand, treating all charges on the same footing (disregarding whether they are bound or free charges), the generalized Ampere's equation (or *Maxwell-Ampère equation*) is:

$$\oint_C B.d\ell = \iint_S \left(\mu_0 J + \mu_0 \varepsilon_0 \frac{\partial}{\partial t} E \right).dA$$

(integral form), or (differential form):

$$\nabla \times B = \left(\mu_0 J + \mu_0 \varepsilon_0 \frac{\partial}{\partial t} E \right),$$

where J includes magnetization current density as well as conduction and polarization current densities. That is, the current density on the right side of the Ampère-Maxwell equation is:

$$J_f + J_D + J_M = J_f + J_P + J_M + \varepsilon_0 \frac{\partial E}{\partial t} = J + \varepsilon_0 \frac{\partial E}{\partial t},$$

where current density J_D is the *displacement current*, and J is the current density contribution actually due to movement of charges, both free and bound. Because $\nabla \cdot D = \rho$, the charge continuity issue with Ampere's original formulation is no longer a problem. Because of the term in $\varepsilon_0 \partial E/\partial t$, wave propagation in free space now is possible. With the addition of the displacement current, Maxwell was able to hypothesize (correctly) that light was a form of electromagnetic wave.

Ampere's Law in cgs Units

In cgs units, the integral form of the equation, including Maxwell's correction, reads;

$$\oint_D B.d\ell = \frac{1}{C}\iint_S \left(4\pi J + \frac{\partial E}{\partial}\right).dS$$

Where *c* is the speed of light.

The differential form of the equation (again, including Maxwell's correction) is

$$\nabla \times B = \frac{1}{C}\left(4\pi J + \frac{\partial E}{\partial t}\right)$$

BIOT-SAVART LAW

The Biot–Savart Law is an equation in electromagnetism that describes the magnetic field B generated by an electric current. The vector field B depends on the magnitude, direction, length, and proximity of the electric current, and also on a fundamental constant called the magnetic constant. The law is valid in the magnetostatic approximation, and results in a B field consistent with both Ampere's circuital law and Gauss's law for magnetism.

The Biot–Savart law is used to compute the magnetic field generated by a *steady current*, i.e. a continual flow of charges, for example through a wire, which is constant in time and in which charge is neither building up nor depleting at any point. The equation is as follows:

$$dB = \frac{\mu_0}{4\pi}\frac{Idl \times \hat{r}}{r^2}$$

(in SI units), where

I is the current, *dl* is a vector, whose magnitude is the length of the differential element of the wire, and whose direction is the direction of conventional current, dB is the differential contribution to the magnetic field resulting from this differential element of wire, μ_0 is the magnetic constant, $\hat{r}$ is the displacement unit vector in the direction pointing from the wire element towards the point at which the field is being computed, r is the distance from the wire element to the point at which the field is being computed, the symbols in boldface denote vector quantities.

To apply the equation, you choose a point in space at which you want to compute the magnetic field. Holding that point fixed, you integrate over the path of the current(s) to find the total magnetic field at that point. The application of this law implicitly relies on the superposition principle for magnetic fields, i.e. the fact that the magnetic field is a vector sum of the field created by each infinitesimal of the wire individually. Another form of the equation is:

$$dB = \frac{\mu_0}{4\pi} \frac{Idl \times r}{r^3}$$

Here, r is the full displacement vector instead of a unit vector and it has distance r^3 in the denominator to compensate. This equation is applied identically to the one above.

The formulations given above work well when the current can be approximated as running through an infinitely-narrow wire. If the current has some thickness, the proper formulation of the Biot-Savart law (again in SI units) is:

$$dB = \frac{\mu_0}{4\pi} \frac{(JdV) \times \hat{r}}{r^3}$$

Where dV is the differential element of volume and J is the current density vector in that volume. The Biot-Savart law is fundamental to magnetostatics, playing a similar role to Coulomb's law in electrostatics.

FORMS OF BIOT–SAVART LAW

General

In the magnetostatic approximation, the magnetic field can be determined if the current density j is known:

$$B = K_m \int \frac{j \times \hat{r}}{r^2} dV$$

where dV is the differential element of volume.

Constant Uniform Current

In the special case of a constant, uniform current I, the magnetic field B is

$$B = K_m I \int \frac{dl \times \hat{r}}{r^2}$$

Point Charge at Constant Velocity

In the case of a charged point particle q moving at a constant, non-relativistic velocity v, then Maxwell's equations give the following expression for the magnetic field:

$$B = K_m \frac{qv \times \hat{r}}{r^2}$$

This equation is also sometimes called the "Biot-Savart law," due to its closely analogous form to the "standard" Biot-Savart law given above. Note that the law is only approximate, with its accuracy decreasing as the particle's velocity approaches c; this happens because the situation is not perfectly approximated by magnetostatics.

This expression can also be rewritten as

$$B = v \times \frac{1}{c^2} E$$

where E is the electric field which the charge would create if it were stationary (as given by Coulomb's law), i.e.

$$E = \frac{1}{4\pi \epsilon_0} \frac{q\hat{r}}{r^2}.$$

The exact, relativistic expression is as follows:

$$B = v \times \frac{1}{c^2} E$$

$$E = \frac{q}{4\pi \epsilon_0} \frac{1 - v^2 / c^2}{(1 - v^2 \sin^2 \theta / c^2)^{3/2}} \frac{\hat{r}}{r^2}$$

Where $\hat{r}$ is the vector pointing from the current (non-retarded) position of the particle to the point at which the field is being measured, and θ is the angle between the velocity vector and $\hat{r}$. The Biot-Savart law can be used in the calculation of magnetic responses even at the atomic or molecular level, e.g. chemical shieldings or magnetic susceptibilities, provided that the current density can be obtained from a quantum mechanical calculation or theory.

MAXWELL'S EQUATIONS

In electromagnetism, Maxwell's equations are a set of four partial differential equations that describe the properties of the electric and magnetic fields and relate them to their sources, charge density and current density. These equations are used to show that light is an electromagnetic wave. Individually, the equations are known as Gauss's law, Gauss's law for magnetism, Faraday's law of induction, and Ampere's law with Maxwell's correction.

These four equations, together with the Lorentz force law are the complete set of laws of classical electromagnetism. The Lorentz force law itself was actually derived by Maxwell under the name of "Equation for Electromotive Force" and was one of an earlier set of eight Maxwell's equations.

Conceptual Description of Maxwell's Equations

This will conceptually describe each of the four Maxwell's equations, and also how they link together to explain the origin of electromagnetic radiation such as light. The exact equations can be found in the subsequent.

- Gauss's law describes how electric charge can create and alter electric fields. In particular, electric fields tend to point away from positive charges, and towards negative charges. Gauss's law is the primary explanation of why opposite charges attract, and like repel: The charges create certain electric fields, which other charges then respond to via an electric force.
- Gauss's law for magnetism states that magnetism is unlike electricity in that there are not distinct "north

pole" and "south pole" *particles* (such particles, which exist in theory only, would be called magnetic monopoles) that attract and repel the way positive and negative charges do. Instead, north poles and south poles necessarily come as pairs (magnetic dipoles). In particular, unlike the electric field which tends to point away from positive charges and towards negative charges, magnetic field lines always come in loops, for example pointing away from the north pole *outside* of a bar magnet but towards it *inside* the magnet.

- Faraday's law of induction describes how a changing magnetic field can create an electric field. This is, for example, the operating principle behind many electric generators: Mechanical force (such as the force of water falling through a hydroelectric dam) spins a huge magnet, and the changing magnetic field creates an electric field which drives electricity through the power grid.
- Ampere's law with Maxwell's correction states that magnetic fields can be generated in two ways: By electrical current (this was the *original* "Ampere's law") and by changing electric fields. The idea that a magnetic field can be induced by a changing electric field follows from the modern concept of displacement current which was introduced to maintain the solenoidal nature of Ampere's law in a vacuum capacitor circuit. This modern displacement current concept has the same mathematical form as Maxwell's original displacement current. Maxwell's original displacement current applies to polarization current in a dielectric medium and it sits adjacent to the modern displacement current in Ampere's law.

Maxwell's correction to Ampere's law was particularly important: In 1864 Maxwell derived the electromagnetic wave equation by linking the displacement current to the time varying electric field that is associated with electromagnetic induction.

The modern extension to displacement current applies in the pure vacuum. This is interpreted as meaning that a changing electric field can produce a magnetic field, *and* vice-versa.

Under this modern interpretation, it follows that, even with no electric charges or currents present, it's possible to have stable, self-perpetuating waves of oscillating electric and magnetic fields, with each field driving the other. The physical parameters of transverse elasticity and density, which Maxwell used to calculate the speed of these electromagnetic waves have now given way to two easily-measurable physical constants called the electric constant and the magnetic constant).

The speed calculated for electromagnetic radiation exactly matches the speed of light; indeed, light *is* one form of electromagnetic radiation (as are X-rays, radio waves, and others). In this way, Maxwell unified the hitherto separate fields of electromagnetism and optics.

General Formulation of Maxwell's Equations

Unlike the equations of mechanics (for example), Maxwell's equations are *not* unchanged in other unit systems. Though the general form remains the same, various definitions get changed and different constants appear at different places. Other than SI (used in engineering), the units commonly used are Gaussian units (based on the cgs system and considered to have some theoretical advantages over SI), Lorentz-Heaviside units (used mainly in particle physics) and Planck units (used in theoretical physics).

Two equivalent, general formulations of Maxwell's equations follow. The first separates bound charge and bound current (which arise in the context of dielectric and/or magnetized materials) from free charge and free current (the more conventional type of charge and current). This separation is useful for calculations involving dielectric or magnetized materials.

The second formulation treats all charge equally, combining free and bound charge into *total* charge (and likewise with current). This is the more fundamental or microscopic point of view, and is particularly useful when no dielectric or magnetic material is present. A proof that these two formulations are mathematically equivalent. Symbols in bold represent vector quantities, whereas symbols in *italics* represent scalar quantities. The definitions of terms used in the two tables of equations are given in another table immediately following.

Table. Formulation in terms of Free Charge and Current

Name	Differential form	Integral Form
Gauss's law:	$\nabla . D = \rho_f$	$\oint_S D.dA = Q_{f,s}$
Gauss's law for magnetism:	$\nabla . B = 0$	$\oint_S B.dA = 0$
Maxwell-Faraday equation(Faraday's law of induction):	$\nabla \times E = -\frac{\partial B}{\partial t}$	$\oint_{\partial S} E.dl = -\frac{\partial \Phi_{B,S}}{\partial t}$
Ampere's circuital law	$\nabla \times H = J_f + \frac{\partial D}{\partial t}$	$\oint_{\partial S} H.dl = I_{f,S} + \frac{\partial \Phi_{D,S}}{\partial t}$

Maxwell's equations are generally applied to *macroscopic averages* of the fields, which vary wildly on a microscopic scale in the vicinity of individual atoms (where they undergo quantum mechanical effects as well).

It is only in this averaged sense that one can define quantities such as the permittivity and permeability of a material.

At the microscopic level, Maxwell's equations, ignoring quantum effects, charges and currents in free space but at this level of detail one must include all charges, even those at an atomic level, generally an intractable problem.

Although James Clerk Maxwell was not the originator of these equations, he nevertheless derived three of them again independently in conjunction with his molecular vortex model of Faraday's "lines of force", along with the full version of Faraday's law of induction. In doing so he made an important addition to Ampere's circuital law.Maxwell also developed Faraday's law of induction into another equation, which used to be listed as a 'Maxwell's equation' but is nowadays known as the Lorentz force law.

Table. Formulation in Terms of Total Charge and Current

Name	Differential form	Integral form
Gauss's law:	$\nabla . E = \frac{\rho}{\varepsilon_0}$	$\oint_S E.dA = \frac{Q_S}{\varepsilon_0}$
Gauss's law for magnetism:	$\nabla . B = 0$	$\oint_S B.dA = 0$
Maxwell-Faraday equation (Faraday's law of induction):	$\nabla \times E = -\frac{\partial B}{\partial t}$	$\oint_{\partial S} E.dl = -\frac{\partial \Phi_{B,S}}{\partial t}$
Ampere's circuital law(with Maxwell's correction):	$\nabla \times B = \mu_0 J + \mu_0 \varepsilon_0 \frac{\partial B}{\partial t}$	$\oint_{\partial S} B.dl = \mu_0 I_S + \mu_0 \varepsilon_0 \frac{\partial \Phi_{E,S}}{\partial t}$

The Term Maxwell's Equations

Controversy has always surrounded the term *Maxwell's equations* concerning the extent to which Maxwell himself was involved in these equations. The term *Maxwell's equations* nowadays applies to a set of four equations that were grouped together as a distinct set in 1884 by Oliver Heaviside, in conjunction with Willard Gibbs.

The importance of Maxwell's role in these equations lies in the correction he made to Ampere's circuital law in his 1861 paper *On Physical Lines of Force*. He added the displacement current term to Ampere's circuital law and this enabled him to derive the electromagnetic wave equation in his later 1865 paper *A Dynamical Theory of the Electromagnetic Field* and demonstrate the fact that light is an electromagnetic wave. This fact was then later confirmed experimentally by Heinrich Hertz in 1887.

The reason that these equations are called *Maxwell's equations* is disputed. Some say that these equations were originally called the Hertz-Heaviside equations but that Einstein for whatever reason later referred to them as the Maxwell-Hertz equations. These equations are based on the works of James Clerk Maxwell, and Heaviside made no secret of the fact that he was working from Maxwell's papers. Heaviside aimed to produce a symmetrical set of equations that were crucial as regards deriving the telegrapher's equations. The net result was a set of four equations, three of which had appeared in substance throughout Maxwell's previous papers, in particular Maxwell's 1861 paper *On Physical Lines of Force* and 1865 paper *A Dynamical Theory of the Electromagnetic Field*. The fourth was a partial time derivative version of Faraday's law of induction that doesn't include motionally induced EMF.

Of Heaviside's equations, the most important in deriving the telegrapher's equations was the version of Ampere's circuital law that had been amended by Maxwell in this 1861 paper to include what is termed the displacement current.

Maxwell's On Physical Lines of Force

Three of Heaviside's four equations appeared throughout Maxwell's 1861 paper *On Physical Lines of Force*:

- At equation of Maxwell's 1861 paper we see $\nabla \times B$ = 0.
- At equation Ampere's circuital law with Maxwell's correction. It is this correction called displacement current which is the most significant aspect of Maxwell's work in electromagnetism as it enabled him to later derive the electromagnetic wave equation in his 1865 paper A Dynamical Theory of the Electromagnetic Field, and hence show that light is an electromagnetic wave. It is therefore this aspect

of Maxwell's work which gives Heaviside's equations their full significance. (Interestingly, Kirchhoff derived the telegrapher's equations in 1857 without using displacement current. But he did use Poisson's equation and the equation of continuity which are the mathematical ingredients of the displacement current. Nevertheless, Kirchhoff believed his equations to be applicable only inside an electric wire and so he is not credited with having discovered that light is an electromagnetic wave).

- At equation ofGauss's law.
- Heaviside's fourth equation introduced a restricted partial time derivative version of Faraday's law of induction. It is important however to note that Heaviside's partial time derivative notation, as opposed to the total time derivative notation used by Maxwell at equations, resulted in the loss of the $v \times B$ term that appeared in Maxwell's equation. Nowadays, the $v \times B$ term appears in the force law $F = q\ (E + v \times B)$ which sits adjacent to Maxwell's equations and bears the name Lorentz force. The Lorentz Force corresponds in effect to Maxwell's equation but it appeared in this paper when Lorentz was still a young boy.

Maxwell's Dynamical Theory of Electromagnetic Field

Confusion over the term "Maxwell's equations" is exacerbated because it is also sometimes used for a set of eight equations that appeared in Part III of Maxwell's 1865 paper A Dynamical Theory of the Electromagnetic Field, entitled "General Equations of the Electromagnetic Field", a confusion compounded by the writing of six of those eight equations as three separate equations (one for each of the Cartesian axes), resulting in twenty equations in twenty unknowns. (this terminology is not common: Modern references to the term "Maxwell's equations" usually refer to the Heaviside restatements.)

These original eight equations are nearly identical to the

Heaviside versions in substance, but they have some superficial differences. In fact, only one of the Heaviside versions is completely unchanged from these original equations, and that is Gauss's law (Maxwell's equation below).

Another of Heaviside's four equations is an amalgamation of Maxwell's law of total currents with Ampere's circuital law. This amalgamation, which Maxwell himself originally made at equation in his 1861 paper "On Physical Lines of Force" is the one that modifies Ampere's circuital law to include Maxwell's displacement current.

The eight original Maxwell's equations can be written in modern vector notation as follows:

- The law of total currents

$$J_{tot} = J + \frac{\partial D}{\partial t}$$

- The equation of magnetic force

$$\mu H = \nabla \times A$$

- Ampere's circuital law

$$\nabla \times H = J_{tot}$$

- Electromotive force created by convection, induction, and by static electricity.

$$E = \mu v \times H - \frac{\partial A}{\partial t} - \nabla \phi$$

- The electric elasticity equation

$$E = \frac{1}{\epsilon} D$$

- Ohm's law

$$E = \frac{1}{\sigma} J$$

- Gauss's law

$$\nabla . D = \rho$$

- Equation of continuity

$$\nabla . J = -\frac{\partial \rho}{\partial t}$$

Notation:

H : Magnetizing field, which Maxwell called the "magnetic intensity".

J : Electric current density (with J being the total current including displacement current).

D: Displacement field (called the "electric displacement" by Maxwell).

ρ : free charge density (called the "quantity of free electricity" by Maxwell).

A : Magnetic vector potential (called the "angular impulse" by Maxwell).

E : Called the "electromotive force" by Maxwell. The term electromotive force is nowadays used for voltage, but it is clear from the context that Maxwell's meaning corresponded more to the modern term electric field.

Φ : Electric potential (which Maxwell also called "electric potential").

σ : Electrical conductivity (Maxwell called the inverse of conductivity the "specific resistance", what is now called the resistivity).

It is interesting to note the $\mu v \times H$ term that appears in equation D. Equation D is therefore effectively the Lorentz force, similarly to equation. However, Maxwell drops the $\mu v \times H$ term from equation D when he is deriving the electromagnetic wave equation, as he considers the situation only from the rest frame.

BOUND CHARGE AND CURRENT

If an electric field is applied to a dielectric material, each of the molecules responds by forming a microscopic dipole—its atomic nucleus will move a tiny distance in the direction of the field, while its electrons will move a tiny distance in the opposite direction. This is called polarization of the material. The distribution of charge that results from these tiny movements turn out to be identical to having a layer of positive charge on one side of the material, and a layer of negative charge on the other side—a macroscopic separation of charge, even though all of the charges involved are "bound" to a single molecule. This is called bound charge. Likewise, in a magnetized material, there is effectively a "bound current" circulating around the material, despite the fact that no individual charge is travelling a distance larger than a single molecule.

Proof That Two General Formulations are Equivalent

A simple proof is outlined which shows that the two alternate general formulations of Maxwell's equations are mathematically equivalent. The relation between polarization, magnetization, bound charge, and bound current is as follows:

$$\rho_b = -\nabla . P$$

$$J_b = \nabla \times M + \frac{\partial P}{\partial t}$$

$$D = \epsilon_0 E + P$$

$$B = \mu_0(\mathrm{H} + \mathrm{M})$$

$$\rho = \rho_b + \rho_f$$

$$J = J_b + J_f$$

Where P and M are polarization and magnetization, and ρ_b and J_b are bound charge and current, respectively. Plugging in these relations, it can be easily demonstrated that the two formulations of Maxwell's equations are precisely equivalent.

CONSTITUTIVE RELATIONS

In order to apply Maxwell's equations (the formulation in terms of free charge and current, and D and H), it is necessary to specify the relations between D and E, and H and B. These are called constitutive relations, and correspond physically to specifying the response of bound charge and current to the field, or equivalently, how much polarization and magnetization a material acquires in the presence of electromagnetic fields.

Case Without Magnetic or Dielectric Materials

In the absence of magnetic or dielectric materials, the relations are simple:

$$D = \epsilon_0 E, H = B/\mu_0$$

Where ε_0 and μ_0 are two universal constants, called the permittivity of free space and permeability of free space, respectively.

Case of Linear Materials

In a "linear", isotropic, nondispersive, uniform material, the relations are also straightforward:

$$D = \epsilon E, H = B/\mu$$

Where ε and μ are constants (which depend on the material), called the permittivity and permeability, respectively, of the material.

General Case

For real-world materials, the constitutive relations are not simple proportionalities, except approximately. The relations can usually still be written:

$$D = \epsilon E, H = B/\mu$$

but ε and μ are not, in general, simple constants, but rather functions. For example, ε and μ can depend upon:

- The strength of the fields (the case of *nonlinearity,* which occurs when ε and μ are functions of E and B effects),
- The direction of the fields (the case of *anisotropy, birefringence,* or *dichroism;* which occurs when ε and μ are second-rank tensors),
- The frequency with which the fields vary (the case of *dispersion,* which occurs when ε and μ are functions of frequency). If further there are dependencies on:
- The position inside the material (the case of a *nonuniform material,* which occurs when the response of the material varies from point to point within the material, an effect called *spatial dispersion;* for example in a domained structure, heterostructure or a liquid crystal, or principally in any bounded medium),
- The history of the fields (the case of *hysteresis,* which occurs when the repsonse of the material is a function of both present and past values of the fields).

Then the constitutive relations take a more complicated form:

$$D(r, t) = \epsilon_0 E(r, t) + P(r, t)$$

$$H(r, t) = \frac{1}{\mu_0} B(r,t) - M(r,t)$$

$$P(r, t) = \epsilon_0 \int d^3 r' dt' \hat{\chi}_{elec}(r,r',t,t';E)E(r',t')$$

$$M(r, t) = \frac{1}{\mu_0} \int d^3 r' dt' \hat{\chi}_{magn}(r,r',t,t';B)B(r',t')$$

In which the permittivity and permeability functions are replaced by integrals over the more general electric and magnetic susceptibilities.

Maxwell's Equations in Terms of E and B

Substituting in the constitutive relations above, Maxwell's equations in a linear material (differential form only) are:

$$\nabla . E = \frac{\rho f}{\epsilon}$$

$$\nabla . B = 0$$

$$\nabla \times E = -\frac{\partial B}{\partial t}$$

$$\nabla \times B = \mu J_f + \mu \epsilon \frac{\partial E}{\partial t}.$$

These are formally identical to the *general* formulation in terms of *E* and *B* except that the permittivity of free space was replaced with the permittivity of the material, the permeability of free space was replaced with the permeability of the material, and only free charges and currents are included (instead of all charges and currents).

IN VACUUM

Starting with the equations appropriate in the case without dielectric or magnetic materials, and assuming that there is no current or electric charge present in the vacuum, we obtain the Maxwell equations in free space:

$$\nabla . E = 0$$

$$\nabla . B = 0$$

$$\nabla \times E = -\frac{\partial B}{\partial t}$$

$$\nabla \times B = \mu_0 \varepsilon_0 \frac{\partial E}{\partial t}.$$

These equations have a solution in terms of traveling sinusoidal plane waves, with the electric and magnetic field directions orthogonal to one another and the direction of travel, and with the two fields in phase, traveling at the speed

$$c = \frac{1}{\sqrt{\mu_0 \varepsilon_0}}$$

In fact, Maxwell's equations explains specifically how these waves can physically propagate through space. The changing magnetic field creates a changing electric field through Faraday's law. That electric field, in turn, creates a changing magnetic field through Maxwell's correction to Ampere's law. This perpetual cycle allows these waves, known as electromagnetic radiation, to move through space, always at velocity *c*.

Maxwell discovered that this quantity *c* equals the speed of light in vacuum (known from early experiments), and concluded (correctly) that light is a form of electromagnetic radiation.

MAGNETIC MONOPOLES

Maxwell's equations of electromagnetism relate the electric and magnetic fields to the motions of electric charges. The standard form of the equations provide for an electric charge, but posit no magnetic charge. Except for this, the equations are symmetric under interchange of electric and magnetic field. In fact, symmetric equations can be written when all charges are zero, and this is how the wave equation is derived.

Fully symmetric equations can also be written if one allows for the possibility of "magnetic charges" analogous to electric charges. With the inclusion of a variable for these magnetic charges, say ρ_m, there will also be "magnetic current" variable in the equations, $\vec{j}_m$. The extended Maxwell's equations, simplified by non dimensionalization, are as follows:

Name	**Without Magnetic Monopoles**	**With Magnetic Monopoles (Hypothetical)**
Gauss's law:	$\vec{\nabla}.\vec{E} = 4\pi\rho_e$	$\vec{\nabla}.\vec{E} = 4\pi\rho_e$
Gauss's law for magnetism:	$\vec{\nabla}.\vec{B} = 0$	$\vec{\nabla}.\vec{B} = 4\pi\rho_m$
Maxwell-Faraday equation (Faraday's law of induction):	$-\vec{\nabla}\times\vec{E} = \frac{\partial\vec{B}}{\partial t}$	$-\vec{\nabla}\times\vec{E} = \frac{\partial\vec{B}}{\partial t} + 4\pi\vec{j}_m$
Ampere's law (with Maxwell's extension):	$\vec{\nabla}\times\vec{B} = \frac{\partial\vec{E}}{\partial t} + 4\pi\vec{j}_e$	$\vec{\nabla}\times\vec{B} = \frac{\partial\vec{E}}{\partial t} + 4\pi\vec{j}_e$

If magnetic charges do not exist, or if they exist but where they are not present in a region, then the new variables are zero, and the symmetric equations reduce to the conventional equations of electromagnetism such as $\vec{\nabla}.\vec{B} = 0$. Classically, the question is *"Why does the magnetic charge always seem to be zero?"*

MATERIALS AND DYNAMICS

The fields in Maxwell's equations are generated by charges and currents. Conversely, the charges and currents are affected by the fields through the Lorentz force equation:

$$F = q(E + v \times B),$$

where q is the charge on the particle and v is the particle velocity. (It also should be remembered that the Lorentz force is not the *only* force exerted upon charged bodies, which also may be subject to gravitational, nuclear, *etc.* forces.) Therefore, in both classical and quantum physics, the precise dynamics of a system form a set of coupled differential equations, which are almost always too complicated to be solved exactly, even at the level of statistical mechanics. This remark applies to not only the dynamics of free charges and currents (which enter Maxwell's equations directly), but also the dynamics of bound charges and currents, which enter Maxwell's equations through the constitutive equations.

Commonly, real materials are approximated as "continuum" media with bulk properties such as the refractive index, permittivity, permeability, conductivity, and/or various susceptibilities. These lead to the *macroscopic* Maxwell's equations, which are written in terms of free charge/current densities and *D, H, E,* and *B* (rather than *E* and *B* alone) along with the constitutive equations relating these fields. For example, although a real material consists of atoms whose electronic charge densities can be individually polarized by an applied field, for most purposes behaviour at the atomic scale is not relevant and the material is approximated by an overall polarization density related to the applied field by an electric susceptibility.

Continuum approximations of atomic-scale in homogeneities cannot be determined from Maxwell's equations alone. but require some type of quantum mechanical analysis such as quantum field theory as applied to condensed matter physics. Various approximate transport equations have evolved, for example, the Boltzmann equation or the Fokker–Planck equation or the Navier-Stokes equations. Some examples where these equations are applied are magneto-hydrodynamics, fluid dynamics, electrohydro-dynamics, superconductivity, plasma modeling.

An entire physical apparatus for dealing with these matters has developed. A different set of *homogenization methods* (evolving from a tradition in treating materials such as conglomerates and laminates) are based upon approximation of an in homo-geneous material by a homogeneous *effective medium* (valid for excitations with wavelengths much larger than the scale of the inhomogeneity).

Theoretical results have their place, but often require fitting to experiment. Continuum-approximation properties of many real materials rely upon measurement, for example, ellipsometry measurements.

In practice, some materials properties have a negligible impact in particular circumstances, permitting neglect of small effects. For example: optical nonlinearities can be neglected for low field strengths; material dispersion is unimportant where frequency is limited to a narrow bandwidth; material absorption can be neglected for wavelengths where a material is transparent; and metals with finite conductivity often are approximated at microwave or longer wavelengths as perfect metals with infinite conductivity (forming hard barriers with zero skin depth of field penetration).

And, of course, some situations demand that Maxwell's equations and the Lorentz force be combined with other forces that are not electromagnetic. An obvious example is gravity. A more subtle example, which applies where electrical forces are weakened due to charge balance in a solid or a molecule, is the Casimir force from quantum electrodynamics.

The connection of Maxwell's equations to the rest of the physical world is via the fundamental sources of charges and currents and the forces on them, and also by the properties of physical materials.

BOUNDARY CONDITIONS

Although Maxwell's equations apply throughout space and time, practical problems are finite and solutions to Maxwell's equations inside the solution region are joined to the remainder of the universe through boundary conditions and started in time using initial conditions. In some cases, like waveguides or cavity resonators, the solution region is largely isolated from the universe, for example, by metallic walls, and boundary conditions at the walls define the fields with influence of the outside world confined to the input/output ends of the structure.

In other cases, the universe at large sometimes is approximated by an artificial absorbing boundary, or, for example for radiating antennas or communication satellites, these boundary conditions can take the form of *asymptotic limits* imposed upon the solution. In addition, for example in an optical fibre or thin-film optics, the solution region often is broken up into subregions with their own simplified properties, and the solutions in each subregion must be joined to each other across the subregion interfaces using boundary conditions. Following are some links of a general nature concerning boundary value problems: Examples of boundary value problems, Sturm-Liouville theory, Dirichlet boundary condition, Neumann boundary condition, mixed boundary condition, Cauchy boundary condition, Sommerfeld radiation condition. Needless to say, one must choose the boundary conditions appropriate to the problem being solved.

CGS UNITS

The above equations are given in the International System of Units, or SI for short. In a related unit system, called cgs (short for centimeter-gram-second), the equations take the following form:

$$\nabla . D = 4\pi\rho_f$$

$$\nabla . B = 0$$

$$\nabla \times E = -\frac{1}{c}\frac{\partial B}{\partial t}$$

$$\nabla \times H = \frac{1}{c}\frac{\partial D}{\partial t} + \frac{4\pi}{c} J_f$$

Where c is the speed of light in a vacuum for the electromagnetic field in a vacuum, the equations become:

$$\nabla . E = 0$$

$$\nabla . B = 0$$

$$\nabla \times E = -\frac{1}{c}\frac{\partial B}{\partial t}$$

$$\nabla \times B = -\frac{1}{c}\frac{\partial E}{\partial t}$$

In this system of units the relation between displacement field, electric field and polarization density is:

$$D = E + 4\pi P$$

And likewise the relation between magnetic induction, magnetic field and total magnetization is:

$$B = H + 4\pi M$$

In the linear approximation, the electric susceptibility and magnetic susceptibility can be defined so that:

$$P = \chi_c E, M = \chi_m H$$

The susceptibilities are dimensionless numbers in both cgs and SI, they have different values in the two unit systems, by a factor of 4π. The permittivity and permeability are:

$$\varepsilon = 1 + 4\pi\chi_e, \mu = 1 + 4\pi\chi_m$$

so that

$$D = \epsilon E, B = \mu H$$

In vacuum, one has the simple relations $\pi = \mu = 1$, $D = E$, and $B = H$.

The force exerted upon a charged particle by the electric field and magnetic field is given by the Lorentz force equation:

$$F = q\left(E + \frac{v}{c} \times B\right),$$

where q is the charge on the particle and v is the particle velocity. This is slightly different from the SI-unit expression above. For example, here the magnetic field B has the same units as the electric field E.

SPECIAL RELATIVITY

Maxwell's equations have a close relation to special relativity: Not only were Maxwell's equations a crucial part of the historical development of special relativity, but also, special relativity has motivated a compact mathematical formulation Maxwell's equations, in terms of covariant tensors. Maxwell's electromagnetic wave equation only applied in what he believed to be the rest frame of the luminiferous medium because he didn't use the $v \times B$ term of his equation when he derived it. Maxwell's idea of the luminiferous medium was that it comprised of aethereal vortices aligned solenoidally along their rotation axes.

The American scientist A.A. Michelson set out to

determine the velocity of the earth through the luminiferous medium aether using a light wave interferometer that he had invented. When the Michelson-Morley experiment was conducted by Edward Morley and Albert Abraham Michelson in 1887, it produced a null result for the change of the velocity of light due to the Earth's motion through the hypothesized aether. This null result was in line with the theory that was proposed in 1845 by George Stokes which suggested that the aether was entrained with the Earth's orbital motion.

Hendrik Lorentz objected to Stokes' aether drag model and along with George FitzGerald and Joseph Larmor, he suggested another approach. Both Larmor and Lorentz derived the Lorentz transformation as one under which Maxwell's equations were invariant. Poincare analyzed the coordination of moving clocks by exchanging light signals. He also established mathematically the group property of the Lorentz transformation. This culminated in Albert Einstein's revolutionary theory of special relativity, which postulated the absence of any absolute rest frame, dismissed the aether as unnecessary and established the invariance of Maxwell's equations in all inertial frames of reference, in contrast to the famous Newtonian equations for classical mechanics. But the transformations between two different inertial frames had to correspond to Lorentz' equations and not–as former believed–to those of Galileo (called Galilean transformations).

Indeed, Maxwell's equations played a key role in Einstein's famous paper on special relativity; for example, in the opening paragraph of the paper, he motivated his theory by nothing that a description of a conductor moving with respect to a magnet must generate a consistent set of fields irrespective of whether the force is calculated in the rest frame of the magnet or that of the conductor.

General relativity has also had a close relationship with Maxwell's equations. For example, Kaluza and Klein showed in the 1920s that Maxwell's equations can be derived by extending general relativity into five dimensions. This strategy of using higher dimensions to unify different forces continues to be an active area of research in particle physics.

Covariant Formulation of Maxwell's Equations

In special relativity, in order to more clearly express the fact that Maxwell's equations in *vacuum* take the same form in any inertial coordinate system, Maxwell's equations are written in terms of four-vectors and tensors in the "manifestly covariant" form. The purely spatial components of the following are in SI units.

One ingredient in this formulation is the electromagnetic tensor, a rank-2 covariant antisymmetric tensor combining the electric and magnetic fields:

$$F_{\alpha,\beta} = \begin{pmatrix} 0 & \frac{-E_x}{c} & \frac{-E_y}{c} & \frac{-E_z}{c} \\ \frac{E_x}{c} & 0 & B_z & -B_y \\ \frac{E_y}{c} & -B_z & 0 & -B_y \\ \frac{E_z}{c} & B_y & -B_x & 0 \end{pmatrix},$$

and the result of raising its indices.

$$F^{\mu v} \underline{\underline{def}} \, \eta^{\mu\alpha} F_{\alpha\beta} \eta^{\beta v} = \begin{pmatrix} 0 & \frac{-E_x}{c} & \frac{E_y}{c} & \frac{E_z}{c} \\ \frac{-E_x}{c} & 0 & B_z & -B_y \\ \frac{-E_y}{c} & -B_z & 0 & B_y \\ \frac{-E_z}{c} & B_y & -B_x & 0 \end{pmatrix}.$$

The other ingredient is the four-current: $J^{\alpha} = (c\rho, \vec{J})$ where ρ is the charge density and J is the current density. With these ingredients, Maxwell's equations can be written:

$$\mu_0 J^{\beta} = \frac{\partial F^{\beta\alpha}}{\partial x^{\alpha}} \, \underline{\underline{\text{def}}} \, \partial_{\alpha} F^{\beta\alpha} \, \underline{\underline{\text{def}}} \, F^{\beta\alpha}{}_{,\alpha}$$

and

$$0 = \partial_\gamma F_{\alpha\beta} + \partial_\beta F_{\gamma\alpha} + \partial_\alpha F_{\beta\gamma} \stackrel{\text{def}}{=} F_{\alpha\beta,\gamma} + F_{\gamma\alpha,\beta} + F_{\beta\gamma,\alpha}$$

The first tensor equation is an expression of the two inhomogeneous Maxwell's equations, Gauss's law and Ampere's law with Maxwell's correction. The second equation is an expression of the two homogeneous equations, Faraday's law of induction and Gauss's law for magnetism. The second equation is equivalent to

$$0 = \epsilon^{\delta\alpha\beta\gamma} F_{\beta\gamma,\alpha}$$

where $\epsilon^{\alpha\beta\gamma\delta}$ is the contravariant version of the Levi-Civita symbol, and

$$\frac{\partial}{\partial x^\alpha} \stackrel{def}{=} \partial_\alpha \stackrel{def}{=} {}_{,\alpha} \stackrel{def}{=} \left(\frac{1}{c}\frac{\partial}{\partial t}, \nabla\right)$$

is the 4-gradient. In the tensor equations above, repeated indices are summed over according to Einstein summation convention. We have displayed the results in several common notations. Upper and lower components of a vector, v^α and v_α respectively, are interchanged with the fundamental tensor g, e.g., $g = \eta = \text{diag}(-1, +1, +1, +1)$.

Alternative covariant presentations of Maxwell's equations also exist, for example in terms of the four-potential.

Potentials

Maxwell's equations can be written in an alternative form, involving the electric potential (also called scalar potential) and magnetic potential (also called vector potential), as follows. (The following equations are valid in the absence of dielectric and magnetic materials; or if such materials are present, they are valid as long as bound charge and bound current are included in the total charge and current densities.)

First, Gauss's law for magnetism states:

$$\nabla . B = 0.$$

By Helmholtz's theorem, B can be written in terms of a vector field A, called the magnetic potential:

$$B = \nabla \times A.$$

Second, plugging this into Faraday's law, we get:

$$\nabla \times \left(E + \frac{\partial A}{\partial t} \right) = 0.$$

By Helmholtz's theorem, the quantity in parentheses can be written in terms of a scalar function Φ, called the electric potential:

$$E + \frac{\partial A}{\partial t} = -\nabla \Phi.$$

Combining these with the remaining two Maxwell's equations yields the four relations:

$$E = -\nabla \Phi - \frac{\partial A}{\partial t}$$

$$B = \nabla \times A$$

$$\nabla^2 \Phi + \frac{\partial}{\partial t} (\nabla . A) = -\frac{\rho}{\varepsilon_0}$$

$$\left(\nabla^2 A - \frac{1}{c^2} \frac{\partial^2 A}{\partial t^2} \right) - \nabla \left(\nabla . A + \frac{1}{c^2} \frac{\partial \Phi}{\partial t} \right) = -\mu_0 J.$$

These equations, taken together, are as powerful and complete as Maxwell's equations. The problem has been reduced somewhat, as the electric and magnetic fields each have three components which need to be solved for (six components altogether), while the electric and magnetic potentials have only four components altogether. On the other hand, these equations appear more complicated than Maxwell's equations using just the electric and magnetic fields.

In fact, these equations can be simplified a good deal by taking advantage of gauge freedom—i.e., the fact that there are many different choices of A and Φ consistent with a given E and B.

DIFFERENTIAL FORMS

In free space, where $\varepsilon = \varepsilon_0$ and $\mu = \mu_0$ are constant everywhere, Maxwell's equations simplify considerably once the language of differential geometry and differential forms is used. In what follows, cgs units, not SI units are used, however. The electric and magnetic fields are now jointly described by a 2-form F in a 4-dimensional spacetime manifold. Maxwell's equations then reduce to the Bianchi identity

$$dF = 0$$

where d denotes the exterior derivative—a natural coordinate and metric independent differential operator acting on forms—and the source equation

$$d{*}F = J$$

where the (dual) Hodge star operator * is a linear transformation from the space of 2-forms to the space of forms defined by the metric in Minkowski space (in four dimensions even by any metric conformal to this metric), and the fields are in natural units where $1 / 4\pi\varepsilon_0 = 1$. Here, the 3-form J is called the "electric current form" or "current 3-form" satisfying the continuity equation

$$dJ = 0.$$

The current 3-form can be integrated over a 3-dimensional space-time region. The physical interpretation of this integral is the charge in that region if it is spacelike, or the amount of charge that flows through a surface in a certain amount of time if that region is a spacelike surface cross a timelike interval. As the exterior derivative is defined on any manifold, the differential form version of the Bianchi identity makes sense for any 4-dimensional manifold, whereas the source equation is defined if the manifold is oriented and has a Lorentz metric. In particular the differential form version of the Maxwell equations are a convenient and intuitive formulation of the Maxwell equations in general relativity.

In a linear, macroscopic theory, the influence of matter on the electromagnetic field is described through more general linear transformation in the space of 2-forms.

$$C : \Lambda^2 \ni \; F \mapsto G \in \Lambda^{(4-2)}$$

We call the constitutive transformation. The role of this transformation is comparable to the Hodge duality transformation. The Maxwell equations in the presence of matter then become:

$$dF = 0$$
$$dG = J$$

where the current 3-form J still satisfies the continuity equation $dJ = 0$.

When the fields are expressed as linear combinations (of exterior products) of basis forms θ^p,

$$F = \frac{1}{2} F_{pq} \theta^p \wedge \theta^q.$$

the constitutive relation takes the form

$$G_{pq} = C_{pq}^{mn} F_{mn}$$

where the field coefficient functions are antisymmetric in the indices and the constitutive coefficients are antisymmetric in the corresponding pairs. In particular, the Hodge duality transformation leading to the vacuum equations are obtained by taking

$$C_{pq}^{mn} = \frac{1}{2} g^{ma} g^{nb} \in_{abpq} \sqrt{-g}$$

which up to scaling is the only invariant tensor of this type that can be defined with the metric.

In this formulation, electromagnetism generalises immediately to any 4-dimensional oriented manifold or with small adaptations any manifold, requiring not even a metric. Thus the expression of Maxwell's equations in terms of differential forms leads to a further notational and conceptual simplification. Whereas Maxwell's Equations could be written as two tensor equations instead of eight scalar equations, from which the propagation of electromagnetic disturbances and the continuity equation could be derived with a little effort, using differential forms leads to an even simpler derivation of these results.

Conceptual Insight from Formulation

On the conceptual side, from the point of view of physics, this shows that the second and third Maxwell equations should be grouped together, be called the homogeneous ones, and geometric *identities* expressing nothing else than: the *field F* derives from a more "fundamental" *potential A*. While the first and last one as the dynamical *equations of motion*, obtained via the Lagrangian principle of least action, from the "interaction term" $A\ J$ (introduced through gauge covariant derivatives), coupling the field to matter.

Often, the time derivative in the third law motivates calling this equation "dynamical", which is somewhat misleading; in the sense of the preceding analysis, this is rather an artifact of breaking relativistic covariance by choosing a preferred time direction. To have physical degrees of freedom propagated by these field equations, one must include a kinetic term $F * F$ for A; and take into account the non-physical degrees of freedom which can be removed by gauge transformation $A \to A' = A - d\alpha$.

ELECTRODYNAMICS AS CURVATURE OF LINE BUNDLE

An elegant and intuitive way to formulate Maxwell's equations is to use complex line bundles or principal bundles with fibre U. The connection ∇ on the line bundle has a curvature $F = \nabla^2$ which is a two-form that automatically satisfies $dF = 0$ and can be interpreted as a field-strength. If the line bundle is trivial with flat reference connection d we can write $\nabla = d + A$ and $F = dA$ with A the 1-form composed of the electric potential and the magnetic vector potential.

In quantum mechanics, the connection itself is used to define the dynamics of the system. This formulation allows a natural description of the Aharonov-Bohm effect. In this experiment, a static magnetic field runs through a long magnetic wire (e.g., an Fe wire magnetized longitudinally). Outside of this wire the magnetic induction is zero, in contrast to the vector potential, which essentially depends on the magnetic flux through the cross-section of the wire and does not vanish outside. Since there is no electric field either, the Maxwell tensor $F = 0$ throughout the space-time region outside the tube, during the experiment. This means by definition that the connection ∇ is flat there.

However, as mentioned, the connection depends on the magnetic field through the tube since the holonomy along a non-contractible curve encircling the tube is the magnetic flux through the tube in the proper units. This can be detected quantum-mechanically with a double-slit electron diffraction experiment on an electron wave traveling around the tube. The holonomy corresponds to an extra phase shift, which leads to a shift in the diffraction pattern.

CURVED SPACE TIME

Traditional Formulation

Matter and energy generate curvature of spacetime. This is the subject of general relativity. Curvature of spacetime affects electrodynamics. An electromagnetic field having energy and momentum will also generate curvature in spacetime. Maxwell's equations in curved spacetime can be obtained by replacing the derivatives in the equations in flat spacetime with covariant derivatives. The sourced and source-free equations become (cgs units):

$$\frac{4\pi}{c} j^{\beta} = \partial_{\alpha} F^{\alpha\beta} + \Gamma^{\alpha}{}_{\mu\alpha} F^{\alpha\mu} \stackrel{\text{def}}{=} D_{\alpha} F^{\alpha\beta} \stackrel{\text{def}}{=} F^{\alpha\beta}{}_{;\alpha},$$

and

$$0 = \partial_{\gamma} F_{\alpha\beta} + \partial_{\beta} F_{\gamma\alpha} + \partial_{\alpha} F_{\beta\gamma} = D_{\gamma} F_{\alpha\beta} + D_{\beta} F_{\gamma\alpha} + D_{\alpha} F_{\beta\gamma}.$$

Here,

$$\Gamma^{\alpha}{}_{\mu\beta}$$

Is a Christoffel symbol that characterizes the curvature of spacetime and D_{γ} is the covariant derivative.

Formulation in Differential Forms

The formulation of the Maxwell equations in terms of differential forms can be used without change in general relativity. The equivalence of the more traditional general relativistic formulation using the covariant derivative with the differential form formulation can be seen as follows. Choose local coordinates x^α which gives a basis of 1-forms dx^α in every point of the open set where the coordinates are defined. Using this basis and cgs units we define.

- The antisymmetric infinitesimal field tensor $F_{\alpha\beta}$, corresponding to the field 2-form F

$$F := \frac{1}{2} F_{\alpha\beta} dx^\alpha \wedge dx^\beta.$$

- The current-vector infinitesimal 3-form J

$$J := \frac{4\pi}{c} j^\alpha \sqrt{-g}\, \epsilon_{\alpha\beta\gamma\delta}\, x^\beta \wedge dx^\gamma \wedge dx^\delta.$$

Here g is as usual the determinant of the metric tensor $g_{\alpha\beta}$. A small computation that uses the symmetry of the Christoffel symbols (i.e. the torsion-freeness of the Levi Civita connection) and the covariant constantness of the Hodge star operator then shows that in this coordinate neighborhood we have:

- The Bianchi identity

$$dF = 2(\partial_\gamma F_{\alpha\beta} + \partial_\beta F_{\gamma\alpha} + \partial_\alpha F_{\beta\gamma}) dx^\alpha \wedge dx^\beta \wedge dx^\gamma = 0$$

- The source equation

$$d * F = F^{\alpha\beta};_\alpha \sqrt{-g}\, \epsilon_{\beta\gamma\delta\eta}\, dx^\gamma \wedge dx^\delta \wedge dx^\eta = J$$

- The continuity equation

$$dJ = \frac{4\pi}{c} j^\alpha;_\alpha \sqrt{-g}\, \epsilon_{\alpha\beta\gamma\delta}\, dx^\alpha \wedge dx^\beta \wedge dx^\gamma \wedge dx^\delta = 0.$$

Chapter 2

Electromagnetic Radiation

Electromagnetic Radiation, energy waves produced by the oscillation or acceleration of an electric charge. Electromagnetic waves have both electric and magnetic components. Electromagnetic radiation can be arranged in a spectrum that extends from waves of extremely high frequency and short wavelength to extremely low frequency and long wavelength. Visible light is only a small part of the electromagnetic spectrum.

In order of decreasing frequency, the electromagnetic spectrum consists of gamma rays, hard and soft X rays, ultraviolet radiation, visible light, infrared radiation, microwaves, and radio waves.

PROPERTIES OF ELECTROMAGNETIC RADIATION

There are three phenomena through which energy can be transmitted: electromagnetic radiation, conduction, and convection. Unlike conduction and convection, electromagnetic waves need no material medium for transmission. Thus, light and radio waves can travel through interplanetary and interstellar space from the sun and stars to the earth. Regardless of the frequency, wavelength, or method of propagation, electromagnetic waves travel at a speed of 3×10^{10} cm (186,272 mi) per second in a vacuum.

All the components of the electromagnetic spectrum, regardless of frequency, also have in common the typical properties of wave motion, including diffraction and interference. The wavelengths range from millionths of a centimeter to many kilometers. The wavelength and frequency of electromagnetic waves are important in determining heating effect, visibility, penetration, and other characteristics of the electromagnetic radiation.

THEORY OF ELECTROMAGNETIC RADIATION

British physicist James Clerk Maxwell laid out the theory of electromagnetic waves in a series of papers published in the 1860s. He analyzed mathematically the theory of electromagnetic fields and predicted that visible light was an electromagnetic phenomenon. Physicists had known since the early 19th century that light is propagated as a transverse wave (a wave in which the vibrations move in a direction perpendicular to the direction of the advancing wave front).

They assumed, however, that the wave required some material medium for its transmission, so they postulated an extremely diffuse substance, called ether, as the unobservable medium. Maxwell's theory made such an assumption unnecessary, but the ether concept was not abandoned immediately, because it fit in with the Newtonian concept of an absolute space-time frame for the universe.

A famous experiment conducted by the American physicist Albert Abraham Michelson and the American chemist Edward Williams Morley in the late 19th century served to dispel the ether concept and was important in the development of the theory of relativity. This work led to the realization that the speed of electromagnetic radiation in a vacuum is an invariant.

QUANTA OF RADIATION

At the beginning of the 20th century, however, physicists found that the wave theory did not account for all the properties of radiation. In 1900 the German physicist Max Planck demonstrated that the emission and absorption of radiation occur in finite units of energy, known as *quanta*. In 1904, German-born American physicist Albert Einstein was able to explain some puzzling experimental results on the external photoelectric effect by postulating that electromagnetic radiation can behave like a particle. Other phenomena, which occur in the interaction between radiation and matter, can also be explained only by the quantum theory. Thus, modern physicists were forced to recognize that electromagnetic radiation can sometimes behave like a particle, and sometimes behave like a wave. The parallel concept—that matter also exhibits the same duality of having particlelike and wavelike characteristics—was developed in 1923 by the French physicist Louis Victor, Prince de Broglie.

X-RAY

X-Ray, penetrating electromagnetic radiation, having a shorter wavelength than light, and produced by bombarding a target, usually made of tungsten, with high-speed electrons. X-rays were discovered accidentally in 1895 by the German physicist Wilhelm Conrad Roentgen while he was studying cathode rays in a high-voltage, gaseous-discharge tube. Despite the fact that the tube was encased in a black cardboard box, Roentgen noticed that a barium-platinocyanide screen, inadvertently lying nearby, emitted fluorescent light whenever the tube was in operation.

After conducting further experiments, he determined that the fluorescence was caused by invisible radiation of a more penetrating nature than ultraviolet rays. The invisible radiation "X-ray" because of its unknown nature. Subsequently, X-rays were known also as Roentgen rays in his honour.

Nature of X-Rays

X-rays are electromagnetic radiation ranging in wavelength from about 100 A to 0.01 A. The shorter the wavelength of the X-ray, the greater is its energy and its penetrating power. Longer wavelengths, near the ultraviolet-ray band of the electromagnetic spectrum, are known as soft X-rays. The shorter wavelengths, closer to and overlapping the gamma-ray range, are called hard X-rays. A mixture of many different wavelengths is known as "white" X-rays, as opposed to "monochromatic" X-rays, which represent only a single wavelength.

Both light and X-rays are produced by transitions of

electrons that orbit atoms, light by the transitions of outer electrons and X-rays by the transitions of inner electrons. In the case of bremsstrahlung radiation X-rays are produced by the retardation or deflection of free electrons passing through a strong electrical field. Gamma rays, which are identical to X-rays in their effect, are produced by energy transitions within excited nuclei.

X-rays are produced whenever high-velocity electrons strike a material object. Much of the energy of the electrons is lost in heat; the remainder produces X-rays by causing changes in the target's atoms as a result of the impact. The X-rays emitted can have no more energy than the kinetic energy of the electrons that produce them. Moreover, the emitted radiation is not monochromatic but is composed of a wide range of wavelengths with a sharp, lower wavelength limit corresponding to the maximum energy of the bombarding electrons.

This continuous spectrum is referred to by the German name *bremsstrahlung,* which means "braking," or slowing down, radiation, and is independent of the nature of the target. If the emitted X-rays are passed through an X-ray spectrometer, certain distinct lines are found superimposed on the continuous spectrum; these lines, known as the characteristic X-rays, represent wavelengths that depend only on the structure of the target atoms. In other words, a fast-moving electron striking the target can do two things: It can excite X-rays of any energy up to its own energy; or it can excite X-rays of particular energies, dependent on the nature of the target atom.

X-Ray Production

The first X-ray tube was the Crookes tube, a partially evacuated glass bulb containing two electrodes, named after its designer, the British chemist and physicist Sir William Crookes. When an electric current passes through such a tube, the residual gas is ionized and positive ions, striking the cathode, eject electrons from it.

These electrons, in the form of a beam of cathode rays,

bombard the glass walls of the tube and produce X rays. Such tubes produce only soft X rays of low energy. An early improvement in the X-ray tube was the introduction of a curved cathode to focus the beam of electrons on a heavy-metal target, called the anticathode, or anode. This type generates harder rays of shorter wavelengths and of greater energy than those produced by the original Crookes tube, but the operation of such tubes is erratic because the X-ray production depends on the gas pressure within the tube.

The next great improvement was made in 1913 by the American physicist William David Coolidge. The Coolidge tube is highly evacuated and contains a heated filament and a target. It is essentially a thermionic vacuum tube in which the cathode emits electrons because the cathode is heated by an auxiliary current and not because it is struck by ions as in the earlier types of tubes. The electrons emitted from the heated cathode are accelerated by the application of a high voltage across the tube. As the voltage is increased, the minimum wavelength of the radiation decreases.

Most of the X-ray tubes in present-day use are modified Coolidge tubes. The larger and more powerful tubes have water-cooled anticathodes to prevent melting under the impact of the electron bombardment. The widely used shockproof tube is a modification of the Coolidge tube with improved insulation of the envelope (by oil) and grounded power cables. Such devices as the betatron are used to produce extremely hard X-rays, of shorter wavelength than the gamma rays emitted by naturally radioactive elements.

Properties of X-Rays

X-rays affect a photographic emulsion in the same way light does. Absorption of X-radiation by any substance depends upon its density and atomic weight. The lower the atomic weight of the material, the more transparent it is to X-rays of given wavelengths. When the human body is X-rayed, the bones, which are composed of elements of higher atomic weight than the surrounding flesh, absorb the radiation more effectively and therefore cast darker shadows on a photographic plate. Another type of radiation, which is known as neutron radiation and is now used in some types of radiography, produces almost opposite results. Objects that cast dark shadows in an X-ray picture are almost always light in a neutron radiograph.

Fluorescence

X-rays also cause fluorescence in certain materials, such as barium platinocyanide and zinc sulfide. If a screen coated with such fluorescent material is substituted for the photographic films, the structure of opaque objects may be observed directly. This technique is known as fluoroscopy.

Ionization

Another important characteristic of X-rays is their ionizing power, which depends upon their wavelength. The capacity of monochromatic X-rays to ionize is directly proportional to their energy. This property provides a method for measuring the energy of X-rays. When X-rays are passed through an ionization chamber an electric current is produced that is proportional to the energy of the incident beam.

In addition to ionization chambers, more sensitive devices, such as the Geiger-Müller counter and the scintillation counter, can measure the energy of X-rays on the basis of ionization. In addition, the path of X-rays, by virtue of their capacity to ionize, can be made visible in a cloud chamber.

X-Ray Diffraction

X-rays may be diffracted by passage through a crystal or by reflection (scattering) from a crystal, which consists of regular lattices of atoms that serve as fine diffraction gratings. The resulting interference patterns may be photographed and analyzed to determine the wavelength of the incident X-rays or the spacings between the crystal atoms, whichever is the unknown factor.

X-rays may also be diffracted by ruled gratings if the spacings are approximately equal to the wavelengths of the incident X-rays.

Interaction with Matter

In the interaction between matter and X-rays, three mechanisms exist by which X-rays are absorbed; all three mechanisms demonstrate the quantum nature of X-radiation.

Photoelectric Effect

When a quantum of radiation, or a photon, in the X-ray portion of the electromagnetic spectrum strikes an atom, it may impinge on an electron within an inner shell and eject it from the atom. If the photon carries more energy than is necessary to eject the electron, it will transfer its residual energy to the ejected electron in the form of kinetic energy. This phenomenon, called the photoelectric effect, occurs primarily in the absorption of low-energy X-rays.

Compton Effect

The Compton effect, discovered in 1923 by the American physicist and educator Arthur Holly Compton, is an important manifestation of the absorption of X-rays of shorter wavelengths. When a high-energy photon collides with a stationary electron, both particles may be deflected at an angle to the direction of the path of the incident X-ray. The incident photon, having delivered some of its energy to the electron, emerges from the impact with a lower frequency and a correspondingly longer wavelength. These deflections, accompanied by a change of wavelength, are known as Compton scattering.

Pair Production

In the third type of absorption, especially evident when elements of high atomic weight are irradiated with extremely high-energy X-rays, the phenomenon of pair production occurs. When a high-energy photon penetrates the electron shell close to the nucleus, it may create a pair of electrons, one of negative charge and the other positive; a positively charged electron is also known as a positron. This pair production is an example of the conversion of energy into mass. The photon requires at least 1.2 MeV of energy to yield the mass of the pair. If the incident photon possesses more energy than is required for pair production, the excess energy is imparted to the electron pair as kinetic energy. The paths of the two particles are divergent.

Applications of X-Rays

The principal uses of X-radiation are in the field of scientific research, industry, and medicine. The study of X-rays played a vital role in theoretical physics, especially in the development of quantum mechanics. As a research tool, X-rays enabled physicists to confirm experimentally the theories of crystallography. By using X-ray diffraction methods, crystalline substances may be identified and their structure determined.

Virtually all present-day knowledge in this field was either discovered or verified by X-ray analysis. X-ray diffraction methods can also be applied to powdered substances that are not crystalline but that display some regularity of molecular structure. By means of such methods, chemical compounds can be identified and the size of ultramicroscopic particles can be established. Chemical elements and their isotopes may be identified by X-ray spectroscopy, which determines the wavelengths of their characteristic line spectra.

Several elements were discovered by analysis of X-ray spectra.A number of recent applications of X-rays in research are assuming increasing importance. Microradiography, for instance, produces fine-grain images that can be enlarged considerably. Two radiographs can be combined in a projector to produce a three-dimensional image called a stereoradiogram.

Colour-radiography is also used to enhance the detail of X-ray photographs; in this process, differences in the absorption of X-rays by a specimen are shown as different colors. Extremely detailed and analytical information is provided by the electron microprobe, which uses a sharply defined beam of electrons to generate X-rays in an area of specimen as small as 1 micrometer (about 1/25,000 in) square.

Industry

In addition to the research applications of X-rays in physics, chemistry, mineralogy, metallurgy, and biology, X-rays are used in industry as a research tool and for many testing processes. They are valuable in industry as a means of testing objects such as metallic castings without destroying them. X-ray images on photographic plates reveal the presence of flaws, but a disadvantage of such inspection is that the necessary high-powered X-ray equipment is bulky and expensive.

In some instances, therefore, radioisotopes, which emit highly penetrating gamma rays, are used instead of X-ray equipment. These isotope sources can be housed in relatively light, compact, and shielded containers. Cobalt-60 and cesium -137 have been used widely for industrial radiography. Thulium-70 has been used in small, convenient, isotope projectors for some medical and industrial applications.

Many industrial products are inspected routinely by means of X-rays so that defective products may be eliminated at the point of production. Other applications include the detection of fake gems and the detection of smuggled goods in customs examinations. Ultrasoft X-rays are used to determine the authenticity of works of art and for art restoration.

Medicine

X-ray photographs, called radiographs, and fluoroscopy are used extensively in medicine as diagnostic tools. In radiotherapy, X-rays are used to treat certain diseases, notably cancer, by exposing tumors to X-radiation. The use of radiographs for diagnostic purposes was inherent in the penetrating properties of X-rays. Within a few years of their discovery, X-rays were being used to locate foreign bodies, such as bullets, within the human body. With the development of improved X-ray techniques, minute differences in tissues were revealed by radiographs, and many pathological conditions could be diagnosed by means of X-rays. X-rays provided the most important single method of diagnosing tuberculosis when that disease was prevalent. Pictures of the lungs were easy to interpret because the air spaces are more transparent to X-rays than the lung tissues. Various other cavities in the body can be filled artificially with contrasting media, either more transparent or more opaque to X-rays than the surrounding tissue, so that a particular organ is brought more sharply into view. Barium sulfate, which is highly opaque to X-rays, is used for the X-ray examination of the gastrointestinal tract.

Certain opaque compounds are administered either by mouth or by injection into the bloodstream in order to examine the kidneys or the gallbladder. Such dyes can have serious side effects, however, and should be used only after careful consultation. The routine use of X-ray diagnosis has in fact been discouraged—by the American College of Radiology in 1982, for example—as of questionable usefulness.

A recent X-ray device, used without dyes, offers clear views of any part of the anatomy, including soft organ tissues. Called the body scanner, or computerized axial tomography (CAT or CT) scanner, it rotates 180° around a patient's body, sending out a pencil-thin X-ray beam at 160 different points. Crystals positioned at the opposite points of the beam pick up and record the absorption rates of the varying thicknesses of tissue and bone.

These data are then relayed to a computer that turns the information into a picture on a screen. Using the same dosage of radiation as that of the conventional X-ray machine, an entire "slice" of the body is made visible with about 100 times more clarity. The scanner was invented in 1972 by the British electronics engineer Godfrey N. Hounsfield, and was in general use by 1979.

ULTRAVIOLET RADIATION

Ultraviolet Radiation, electromagnetic radiation that has wavelengths in the range between 4000 angstrom units (Å), the wavelength of violet light, and 150 Å, the length of X-rays. Natural ultraviolet radiation is produced principally by the sun. Ultraviolet radiation is produced artificially by electric-arc lamps. Ultraviolet radiation is often divided into three categories based on wavelength, *UV-A, UV-B,* and *UV-C.* In general shorter wavelengths of ultraviolet radiation are more dangerous to living organisms. *UV-A* has a wavelength from 4000 Å to about 3150 Å. *UV-B* occurs at wavelengths from about 3150 Å to about 2800 Å and causes sunburn; prolonged exposure to *UV-B* over many years can cause skin cancer. *UV-C* has wavelengths of about 2800 Å to 150 Å and is used to sterilize surfaces because it kills bacteria and viruses.

The earth's atmosphere protects living organisms from the sun's ultraviolet radiation. If all the ultraviolet radiation produced by the sun were allowed to reach the surface of the earth, most life on earth would probably be destroyed. Fortunately, the ozone layer of the atmosphere absorbs almost all of the short-wavelength ultraviolet radiation, and much of the long-wavelength ultraviolet radiation. However, ultraviolet radiation is not entirely harmful; a large portion of the vitamin *D* that humans and animals need for good health is produced when the human's or animal's skin is irradiated by ultraviolet rays.

When exposed to ultraviolet light, many substances behave differently than when exposed to visible light. For example, when exposed to ultraviolet radiation, certain minerals, dyes, vitamins, natural oils, and other products become *fluorescent*—that is, they appear to glow. Molecules in the substances absorb the invisible ultraviolet light, become energetic, then shed their excess energy by emitting visible light. As another example, ordinary window glass, transparent to visible light, is opaque to a large portion of ultraviolet rays, particularly ultraviolet rays with short wavelengths. Special-formula glass is transparent to the longer ultraviolet wavelengths, and quartz is transparent to the entire naturally occurring range.

SUN

Sun, closest star to Earth. The Sun is a huge mass of hot, glowing gas. The strong gravitational pull of the Sun holds Earth and the other planets in the solar system in orbit. The

Sun's light and heat influence all of the objects in the solar system and allow life to exist on Earth.

The Sun is an average star its size, age, and temperature fall in about the middle of the ranges of these properties for all stars. Astronomers believe that the Sun is about 4.6 billion years old and will keep shining for about another 7 billion years.

For humans, the Sun is beautiful and useful, but also powerful and dangerous. As Earth turns, the Sun rises over the eastern horizon in the morning, passes across the sky during the day, and sets in the west in the evening. This movement of the Sun across the sky marks the passage of time during the day. The Sun's movement can produce spectacular sunrises and sunsets under the right atmospheric conditions. At night, reflected sunlight makes the Moon and planets bright in the night sky.

The Sun provides Earth with vast amounts of energy every day. The oceans and seas store this energy and help keep the temperature of Earth at a level that allows a wide variety of life to exist. Plants use the Sun's energy to make food, and plants provide food for other organisms. The Sun's energy also creates wind in Earth's atmosphere. This wind can be harnessed and used to produce power.

While it lights our day and provides energy for life, sunlight can also be harmful to people. Human skin is sensitive to ultraviolet light emitted from the Sun. Earth's atmosphere blocks much of the harmful light, but sunlight is still strong enough to burn skin under some conditions. Sunburn is one of the most important risk factors in the development of skin cancers, which can be fatal. Sunlight is also very harmful to human eyes. A person should never look directly at the Sun, even with sunglasses or during an eclipse. The Sun influences Earth with more than just light. Particles flowing from the Sun can disrupt Earth's magnetic field, and these disruptions can interfere with electronic communications.

PHYSICAL CHARACTERISTICS OF SUN

The Sun is large and massive compared to the other objects

in the solar system. The Sun's radius (the distance from its centre to its surface) is 695,508 km (432,169 mi), 109 times as large as Earth's radius. If the Sun were hollow, a million Earths could fit inside it. The Sun has a mass of 1.989×10^{27} metric tons. This number is very large. Written out, it would be the digits 1989 followed by 24 zeroes. The Sun is 333,000 times as massive as Earth is. Despite its large mass, the Sun has a lower density, or mass per unit volume, than Earth. The Sun's average density is only 1.409 g/cu cm (1.188 oz/cu in), which is a quarter of the average density of Earth.

The Sun produces an enormous amount of light. It generates 3.83×10^{26} watts of power in the form of light. In comparison, an incandescent lamp emits 60 to 100 watts of power. The temperature of the outer, visible part of the Sun is 5500°C (9900°F).

From Earth the Sun looks small, because it is far away. Its average distance from Earth is 150 million km (93 million mi). Light from the Sun takes about eight minutes to reach Earth. This light is still strong enough when it reaches Earth, however, to damage human eyes when viewed directly. The Sun is much closer to Earth than any other star is. The Sun's nearest stellar neighbour, Proxima Centauri (part of the triple star Alpha Centauri), is 4.3 light-years from our solar system, meaning light from Proxima Centauri takes 4.3 years to reach the Sun. The Sun is so much closer to Earth than all other stars are that the intense light of the Sun keeps us from seeing any other stars during the day.

Importance to Earth

Earth would not have any life on it without the Sun's energy, which reaches Earth in the form of heat and light. This energy warms our days and illuminates our world. Green plants absorb sunlight and convert it to food, which these plants then use to live and grow. In this process, the plants give off the oxygen that animals breathe. Animals eat these plants for nourishment. All plant and animal life relies on the Sun's presence.

The Sun also provides—directly or indirectly—much of

the energy on Earth that people use for fuel. Devices called solar cells turn sunlight into electricity. Sunlight can heat a gas or liquid, which can then be circulated through a building to heat the building. The energy stored in fossil fuels originally came from the Sun. Ancient plants used sunlight as fuel to grow. Animals consumed these plants.

The plants and animals stored the energy of sunlight in the organic material that composed them. When the ancient plants and animals died and decayed, this organic material was buried and gradually turned into the petroleum, coal, and natural gas people use today. The Sun's energy produces the winds and the movements of water that people harness to produce electricity.

The Sun heats Earth's oceans and land, which in turn heat the air and make it circulate in the atmosphere as wind. The Sun fuels Earth's water cycle, evaporating water from the oceans, seas, and lakes. This water returns to the ground in the form of precipitation, flowing back to the oceans through the ground and in rivers. The energy of water's motion in rivers can be harnessed with dams.

Role in the Solar System

The Sun's gravitational pull holds the solar system together. The planets, asteroids, comets, and dust that make up our solar system are strongly attracted to the Sun's huge mass. This gravitational attraction keeps these bodies in orbit around the Sun. The Sun also influences the solar system with its diffuse outer atmosphere, which expands outward in all directions. This expanding atmosphere fills the solar system with a constant flow of tiny, fast, electrically charged particles. This flow is called the solar wind.

The region through which the solar wind blows is called the heliosphere. Estimates vary about the extent of the heliosphere, ranging from about 86 to about 100 times the distance between Earth and the Sun. Interstellar winds may give the heliosphere an egg shape. The solar wind spreads out as it leaves the Sun. The point at which the solar wind is so diffuse that it stops having an effect on its surroundings is called the heliopause. The heliopause marks the outer edge of the solar system. Within the heliosphere, the Sun provides most of the heat and light that are present, and the particles in the solar wind interact with the planets and satellites in the solar system.

The solar wind causes auroras—displays of colored light—in the atmosphere of Earth's polar regions. The solar wind also carries remnants of the Sun's magnetic field, which affect the magnetic fields of the planets and larger satellites. The solar wind pushes the planets' magnetic fields away from the Sun, turning them into elongated, windsock shapes.

THE SUN'S ENERGY

The Sun produces an amazing amount of light and heat through nuclear reactions (Nuclear Energy). The process that produces the Sun's energy is called nuclear fusion. In nuclear fusion, two atoms come together to produce a heavier atom. Fusion reactions release energy and tiny elementary particles.

Scale of the Sun's Energy

In just one second the Sun emits more energy than humans have used in the last 10,000 years. The Sun has been shining relatively steadily for 4.6 billion years. Until the early 20th century, humans did not know of any process that could explain the energy production of the Sun. Even if a fire, such as those that occur on Earth, were as large as the Sun, the fire would consume the mass of the Sun in a few thousand years.

Scientists now know that the Sun is mainly composed of hydrogen, the lightest and most abundant element in the universe. The Sun contains an enormous amount of hydrogen, however, which makes the Sun very massive. All matter inside the Sun is gravitationally attracted to all the other matter in the Sun, and this attraction tends to pull the Sun's mass together. This inward pull creates high pressures and temperatures inside the Sun. The centre is so violent and hot that collisions between atoms break the hydrogen atoms apart into their subatomic ingredients.

A hydrogen atom is made up of a nucleus that contains a

positively charged proton, and a negatively charged electron that orbits the nucleus. In the Sun, collisions separate the electron from the nucleus, freeing each to move about the solar interior. The positively charged nuclei, or protons, are called ions. A gas in which particles are ionized, or have electric charges, is called plasma. Scientists often consider plasma, such as the material inside the Sun, to be a fourth state of matter—the three more familiar states of matter are gas, liquid, and solid.

Nuclear Fusion in the Core

The separation of hydrogen nuclei from their electrons makes nuclear fusion possible at the Sun's core, producing the Sun's light and heat. With their electrons gone, hydrogen nuclei (protons) can be packed much more tightly than complete atoms. At great depths inside the Sun, the pressure of overlying material is enormous, the protons are squeezed tightly together, and the material is very hot and densely concentrated.

At the Sun's centre, the temperature is 15.6 million degrees C (28.1 million degrees F), and the density is more than 13 times that of solid lead. This is hot and dense enough to make the nuclei fuse together. Outside the solar core, where the overlying weight and compression are less, the gas is cooler and thinner, and nuclear fusion cannot occur.

The nuclear fusion reaction that powers the Sun involves four protons that fuse together to make one nucleus of helium. Two of the original protons become neutrons (electrically neutral particles about the same size as protons). The result is a helium nucleus, containing two protons and two neutrons. The helium nucleus is slightly less massive (by a mere 0.7 per cent) than the four protons that combine to make it. The fusion reaction turns the missing mass into energy, and this energy powers the Sun.

The relationship between energy and the missing matter was explained in 1905 by German-born American physicist Albert Einstein. The mass loss, m, during the transformation of four protons into one helium nucleus, supplies an energy,

E, according to the relation $E = mc^2$, where *c* is the speed of light. The speed of light is a constant number equal to 3×10^8 m/s (1×10^9 ft/s).

Every second, fusion reactions convert about 700 million metric tons of hydrogen into helium within the Sun's energy-generating core. In doing so, about 5 million metric tons of this matter become energy. This energy leaves the Sun as radiation, and the part of this radiation that constitutes visible light is what makes the Sun shine.

The rate of nuclear reactions in the Sun is relatively low, because protons repel each other. This repulsion often prevents them from getting close enough to each other to fuse. Protons push each other away because they have the same electrical charge. The particles must overcome this repulsion in order to fuse together. Only a tiny fraction of the protons inside the Sun are moving fast enough to overpower this repulsive electrical force. The nuclei that are moving fast enough can get very close together, and a force called the strong nuclear force takes over. The strong nuclear force is, as its name implies, very powerful, but only over very short distances. It pulls the nuclei together and holds them together. In this way, nuclear reactions proceed at a relatively slow pace inside the Sun. If the pace were much quicker, the Sun would explode like a giant hydrogen bomb.

The Proton-Proton Chain

Four protons do not combine directly to form a helium nucleus, since the protons are constantly moving and are almost never in the same place at the same time. The electrical repulsion between four protons is too great to overcome, even if the four protons happen to come together at an appropriate speed at the same time. Instead, the protons come together in a series of steps to form a helium nucleus, and these steps are called the proton-proton chain.

In the first step of the proton-proton chain, two exceptionally fast protons meet head on and merge into each other, tunneling through the electrical barrier between them. The two protons combine, with most of their mass forming a deuteron, the nucleus of a heavy form of hydrogen known as deuterium. A deuteron contains one proton and one neutron, so one of the protons must become a neutron in this step. The conversion of a proton to a neutron releases a much smaller particle called a neutrino. There are several types of neutrinos—the type that the proton-proton chain produces is called an electron neutrino. The reaction also creates a positron, a positively charged particle the size of an electron. The symbolic representation of the first step of the proton-proton chain is

$$p + p \rightarrow {}^{2}D + e^{+} + v_{e}$$

where p represents the protons, ${}^{2}D$ represents deuterium, e^{+} represents the positron, and v_e represents the electron neutrino.

In the second step of the chain, the deuteron collides with another proton to form a nucleus of light helium, which has two protons and one neutron. Less energy is needed to maintain a light helium nucleus than is needed to maintain a deuteron and a proton separately. The extra energy is released as a photon, or a packet of light energy. In symbolic terms, the second step is

$${}^{2}D + p \rightarrow {}^{3}\mathrm{He} + \gamma$$

where ^{3}He is light helium and g represents a photon.

In the final step of the proton-proton chain, two light helium nuclei meet and fuse together to form a nucleus of normal heavy helium, which has two protons and two neutrons. This reaction also releases two unattached hydrogen nuclei that return to the solar gas. In symbolic terms, the third step is

$${}^{3}\mathrm{He} + {}^{3}\mathrm{He} \rightarrow {}^{4}\mathrm{He} + 2\mathrm{p}$$

where ^{4}He represents a normal helium nucleus with two protons and two neutrons.

The positron created in the first step of the chain eventually collides with a free electron. The positron and the electron are opposite particles—the positron is the antimatter equivalent of the electron. When the positron and the electron collide, they annihilate each other, releasing energy. The electron and the positron disappear, their mass transformed into two photons:

$$e^+ + e^- \rightarrow 2\gamma$$

where e^- represents the electron. The net result of the proton-proton chain is the transformation of four hydrogen nuclei into a helium nucleus (with two protons and two neutrons), two neutrinos, and six photons:

$$4p \rightarrow {}^4\text{He} + 2v_e + 6\gamma.$$

Solar Neutrinos

The conversion of two protons into two neutrons in the proton-proton chain produces two tiny, elusive, fast-moving neutral particles called neutrinos. Nuclear reactions in the Sun's central furnace create prodigious quantities of neutrinos. Every second the Sun releases 2×10^{38} neutrinos, and every second an estimated 70 billion of these solar neutrinos pass through every square centimeter of Earth that is facing the Sun.

Neutrinos move at the velocity of light, have no electrical charge, and have so little mass that scientists are not sure that neutrinos have any mass at all. The ghostlike neutrinos therefore travel almost unimpeded through the Sun, Earth, and nearly any amount of matter. Scientists can snag small numbers of neutrinos in massive underground detectors called neutrino telescopes. These telescopes are placed so deep underground that only neutrinos can reach them. Scientists using these telescopes have detected solar neutrinos, confirming that the Sun is indeed powered by nuclear fusion.

The number of neutrinos detected by these telescopes, however, is only one-third to one-half of the total number of neutrinos predicted to exist by the theory of solar neutrino production. This discrepancy between the number of detected neutrinos and the number predicted is known as the solar neutrino problem. There are two possible explanations—scientists might not understand exactly how the Sun produces its energy, or they could have an incomplete knowledge of neutrinos.

Astronomers are convinced that their models of the Sun are correct and that their predictions for the expected amount of solar neutrinos are therefore correct. Studies of the interior of the Sun substantiate the current models of how the Sun produces its energy, so most scientists agree that the problem lies in their understanding of neutrinos.

The theory scientists favor to explain the problem is that neutrinos from the Sun change on their way to Earth. Scientists know of at least three types of neutrinos. Nuclear fusion reactions in the Sun produce a type of neutrino called an electron neutrino. The other two proven types of neutrinos are called muon neutrinos and tau neutrinos.

Most neutrino telescopes, especially those devoted to solar research, can only detect electron neutrinos. In the 1990s studies of muon neutrinos (produced by reactions between particles called cosmic rays and Earth's atmosphere) showed that muon neutrinos might change into tau neutrinos. Research conducted since the late 1990s indicates that electron neutrinos from the Sun may also change into another type of neutrino. This change would mean the electron neutrino detectors miss many of the Sun's neutrinos.

THE SUN'S SPECTRUM

Sunlight appears yellowish, but it is actually a combination of a rainbow of colors. Scientists use special instruments called spectrographs to separate sunlight out into its different colors. These instruments do the same thing that water molecules in the atmosphere do when the molecules produce a rainbow. Each colour corresponds to a different wavelength of light. Red has the longest wavelength of visible light, and violet has the shortest. The range of wavelengths of sunlight and the intensity at each wavelength are called the Sun's spectrum. The study of the spectra of the Sun and other objects or materials is called spectroscopy.

When sunlight is spread out like a rainbow in the Sun's spectrum, many dark gaps separate one colour from another in the row of colors. These gaps are called absorption lines. Each absorption line is created when sunlight passes through the gases in the Sun's photosphere. Atoms and ions of each element in the gas absorb light at certain wavelengths, creating dark gaps in the Sun's spectrum.

The dark absorption lines in the spectra of the Sun and

other stars fingerprint the ingredients of these stars. Each chemical element produces a unique set of lines, and the presence of these lines shows that a particular element is present in the stellar photosphere. Darker absorption lines indicate greater absorption and therefore larger amounts of the element. Absorption lines in the Sun's spectrum show that hydrogen is by far the most abundant element in the Sun. Other prominent absorption lines are produced by helium, sodium, calcium, and iron. Altogether, 92.1 per cent of the atoms in the Sun are hydrogen atoms, 7.8 per cent are helium atoms, and the other, heavier elements—sodium, calcium, iron, and other elements—make up only 0.1 per cent of the atoms in the Sun. The Sun's absorption lines are called Fraunhofer lines, named after German physicist Joseph von Fraunhofer, who cataloged them in the 1800s. The most common

Fraunhofer lines are listed below, by the letter Fraunhofer gave them, the colour that they block, and the element that causes them.

A (extreme red)	Made by terrestrial oxygen
B (red)	Made by terrestrial oxygen
C (red)	Made by solar hydrogen
D_1 (yellow)	Made by solar sodium
D_2 (yellow)	Made by solar sodium
E (green)	Made by solar iron
F (blue)	Made by solar hydrogen
G (violet)	Made by solar iron and calcium group
H (extreme violet)	Made by solar calcium

The Fraunhofer lines designated *A* and *B* actually have nothing to do with the composition of the Sun. They only appear on spectra gathered within Earth's atmosphere. Earth's atmosphere absorbs sunlight at the wavelengths of the *A* and *B* Fraunhofer lines, creating dark lines on the Sun's spectrum. A spectrum gathered above Earth's atmosphere would not have these lines.

INFRARED RADIATION

Infrared Radiation, emission of energy as electromagnetic

waves in the portion of the spectrum just beyond the limit of the red portion of visible radiation. The wavelengths of infrared radiation are shorter than those of radio waves and longer than those of light waves. They range between approximately 10^{-6} and 10^{-3} (about 0.0004 and 0.04 in). Infrared radiation may be detected as heat, and instruments such as bolometers are used to detect it.

Infrared radiation is used to obtain pictures of distant objects obscured by atmospheric haze, because visible light is scattered by haze but infrared radiation is not. The detection of infrared radiation is used by astronomers to observe stars and nebulas that are invisible in ordinary light or that emit radiation in the infrared portion of the spectrum.

An opaque filter that admits only infrared radiation is used for very precise infrared photographs, but an ordinary orange or light-red filter, which will absorb blue and violet light, is usually sufficient for most infrared pictures. Developed about 1880, infrared photography has today become an important diagnostic tool in medical science as well as in agriculture and industry. Use of infrared techniques reveals pathogenic conditions that are not visible to the eye or recorded on X-ray plates.

Remote sensing by means of aerial and orbital infrared photography has been used to monitor crop conditions and insect and disease damage to large agricultural areas, and to locate mineral deposits. In industry, infrared spectroscopy forms an increasingly important part of metal and alloy research, and infrared photography is used to monitor the quality of products. Infrared devices such as those used during World War II enable sharpshooters to see their targets in total visual darkness. These instruments consist essentially of an infrared lamp that sends out a beam of infrared radiation, often referred to as black light, and a telescope receiver that picks up returned radiation from the object and converts it to a visible image.

Spectrum

Spectrum, rainbowlike series of colors, in the order violet, blue, green, yellow, orange, and red, produced by splitting a composite light, such as white light, into its component colors. Indigo was formerly recognized as a distinct spectral colour. The rainbow is a natural spectrum, produced by meteorological phenomena. A similar effect can be produced by passing sunlight through a glass prism. The first correct explanation of the phenomenon was advanced in 1666 by the English mathematician and physicist Sir Isaac Newton.

When a ray of light passes from one transparent medium, such as air, into another, such as glass or water, it is bent; upon reemerging into the air, it is bent again. This bending is called refraction; the amount of refraction depends on the wavelength of the light. Violet light, for example, is bent more than red light in passing from air to glass or from glass to air. A mixture of red and violet light is thus dispersed into the two colors when it passes through a wedge-shaped glass prism.

A device for producing and observing a spectrum visually is called a spectroscope; a device for observing and recording a spectrum photographically is called a spectrograph; a device for measuring the brightness of the various portions of spectra is called a spectrophotometer; and the science of using spectroscopes, spectrographs, and spectrophotometers to study spectra is called spectroscopy. For extremely accurate spectroscopic measurements, an interferometer is used. During the 19th century, scientists discovered that beyond the violet end of the spectrum, radiations could be detected that were invisible to the human eye but that had marked photochemical action; these radiations were termed ultraviolet.

Similarly, beyond the red end of the spectrum, infrared radiations were detected that, although invisible, transmitted energy, as shown by their ability to raise the temperature of a thermometer. The definition of spectrum was then revised to include these invisible radiations, and has since been extended to include radio waves beyond the infrared, and X rays and gamma rays beyond the ultraviolet.

The term *spectrum* is often loosely applied today to any orderly array produced by analysis of a complex phenomenon. A complex sound such as noise, for example, may be analyzed into an audio spectrum of pure tones of various pitches. Similarly, a complex mixture of elements or isotopes of different atomic weights can be separated into an orderly sequence called a mass spectrum in order of their atomic weights.

Spectroscopy has not only provided an important and sensitive method of chemical analysis, but has also been the chief tool for discoveries in the apparently unrelated fields of astrophysics and atomic theory. In general, changes in motions of the outer electrons of atoms produce spectra in the visible, infrared, and ultraviolet regions. Changes in motions of the inner electrons of heavy atoms produce X-ray spectra. Changes in the configurations of the nucleus of an atom produce gamma-ray spectra. Changes in the configurations of molecules produce visible and infrared spectra.

Different colors of light are similar in consisting of electromagnetic radiations that travel at a speed of approximately 300,000 km per sec (about 186,000 mi per sec). They differ in having varying frequencies and wavelengths, the frequency being equal to the speed of light divided by wavelength. Two rays of light having the same wavelength also have the same frequency and the same colour.

The wavelength of light is so small that it is conveniently expressed in nanometers (nm), which are equal to one-billionth of a meter. The wavelength of violet light varies from about 400 to 450 nm, and of red light from about 620 to 760 nm, or from about 0.000016 to 0.000018 in for violet, and from 0.000025 to 0.000030 in for red.

LIGHT

Light, form of energy visible to the human eye that is radiated by moving charged particles. Light from the Sun provides the energy needed for plant growth. Plants convert the energy in sunlight into storable chemical form through a process called photosynthesis. Petroleum, coal, and natural gas are the remains of plants that lived millions of years ago, and the energy these fuels release when they burn is the chemical energy converted from sunlight. When animals digest the plants and animals they eat, they also release energy stored by photosynthesis. Scientists have learned through experimentation that light behaves like a particle at times and like a wave at other times. The particle-like features are called photons. Photons are different from particles of matter in that they have no mass and always move at the constant speed of about 300,000 km/sec (186,000 mi/sec) when they are in a vacuum. When light diffracts, or bends slightly as it passes around a corner, it shows wavelike behaviour. The waves associated with light are called electromagnetic waves because they consist of changing electric and magnetic fields.

THE NATURE OF LIGHT

To understand the nature of light and how it is normally created, it is necessary to study matter at its atomic level. Atoms are the building blocks of matter, and the motion of one of their constituents, the electron, leads to the emission of light in most sources.

Light Emission

Light can be emitted, or radiated, by electrons circling the nucleus of their atom. Electrons can circle atoms only in certain patterns called orbitals, and electrons have a specific amount of energy in each orbital. The amount of energy needed for each orbital is called an energy level of the atom. Electrons that circle close to the nucleus have less energy than electrons in orbitals farther from the nucleus. If the electron is in the lowest energy level, then no radiation occurs despite the motion of the electron. If an electron in a lower energy level gains some energy, it must jump to a higher level, and the atom is said to be excited. The motion of the excited electron causes it to lose energy, and it falls back to a lower level. The energy the electron releases is equal to the difference between the higher and lower energy levels. The electron may emit this quantum of energy in the form of a photon.

Each atom has a unique set of energy levels, and the energies of the corresponding photons it can emit make up what is called the atom's spectrum. This spectrum is like a fingerprint by which the atom can be identified. The process of identifying a substance from its spectrum is called spectroscopy. The laws that describe the orbitals and energy levels of atoms are the laws of quantum theory. They were invented in the 1920s specifically to account for the radiation of light and the sizes of atoms.

Electromagnetic Waves

The waves that accompany light are made up of oscillating, or vibrating, electric and magnetic fields, which are force fields that surround charged particles and influence other charged particles in their vicinity. These electric and magnetic fields change strength and direction at right angles, or perpendicularly, to each other in a plane (vertically and horizontally for instance). The electromagnetic wave formed by these fields travels in a direction perpendicular to the field's strength (coming out of the plane). The relationship between the fields and the wave formed can be understood by imagining a wave in a taut rope. Grasping the rope and moving it up and down simulates the action of a moving charge upon the electric field. It creates a wave that travels along the rope in a direction that is perpendicular to the initial up and down movement.

Because electromagnetic waves are transverse—that is, the vibration that creates them is perpendicular to the direction in which they travel, they are similar to waves on a rope or waves traveling on the surface of water. Unlike these waves, however, which require a rope or water, light does not need a medium, or substance, through which to travel. Light from the Sun and distant stars reaches Earth by traveling through the vacuum of space.

The waves associated with natural sources of light are irregular, like the water waves in a busy harbor. Scientists think of such waves as being made up of many smooth waves, where the motion is regular and the wave stretches out indefinitely with regularly spaced peaks and valleys. Such regular waves are called monochromatic because they correspond to a single colour of light.

Wavelength, Frequency, and Amplitude

The wavelength of a monochromatic wave is the distance between two consecutive wave peaks. Wavelengths of visible light can be measured in meters or in nanometers (nm), which are one-billionth of a meter (or about 0.4 ten-millionths of an inch). Frequency corresponds to the number of wavelengths that pass by a certain point in space in a given amount of time. This value is usually measured in cycles per second, or hertz (Hz).

All electromagnetic waves travel at the same speed, so in one second, more short waves will pass by a point in space than will long waves. This means that shorter waves have a higher frequency than longer waves. The relationship between wavelength, speed, and frequency is expressed by the equation: wave speed equals wavelength times frequency, or

$$c = \text{l}f$$

Where *c* is the speed of a light wave in m/sec (3×10^8 m/sec in a vacuum), *l* is the wavelength in meters, and *f* is the wave's frequency in Hz.

The amplitude of an electromagnetic wave is the height of the wave, measured from a point midway between a peak and a trough to the peak of the wave. This height corresponds to the maximum strength of the electric and magnetic fields and to the number of photons in the light.

Electromagnetic Spectrum

The electromagnetic spectrum refers to the entire range of frequencies or wavelengths of electromagnetic waves. Light traditionally refers to the range of frequencies that can be seen by humans. The frequencies of these waves are very high, about one-half to three-quarters of a million billion (5×10^{14} to 7.5×10^{14}) Hz. Their wavelengths range from 400 to 700 nm. X rays have wavelengths ranging from several thousandths of a nanometer to several nanometers, and radio waves have wavelengths ranging from several meters to several thousand meters.

Waves with frequencies a little lower than the range of human vision (and with wavelengths correspondingly longer) are called infrared. Waves with frequencies a little higher and wavelengths shorter than human eyes can see are called ultraviolet. About half the energy of sunlight at Earth's surface is visible electromagnetic waves, about 3 per cent is ultraviolet, and the rest is infrared.

Each different frequency or wavelength of visible light causes our eye to see a slightly different colour. The longest wavelength we can see is deep red at about 700 nm. The shortest wavelength humans can detect is deep blue or violet at about 400 nm. Most light sources do not radiate monochromatic light. What we call white light, such as light from the Sun, is a mixture of all the colors in the visible spectrum, with some represented more strongly than others. Human eyes respond best to green light at 550 nm, which is also approximately the brightest colour in sunlight at Earth's surface.

B3 Polarization

Polarization refers to the direction of the electric field in an electromagnetic wave. A wave whose electric field is oscillating in the vertical direction is said to be polarized in the vertical direction. The photons of such a wave would interact with matter differently than the photons of a wave polarized in the horizontal direction. The electric field in light waves from the Sun vibrates in all directions, so direct sunlight is called unpolarized. Sunlight reflected from a surface is partially polarized parallel to the surface. Polaroid sunglasses block light that is horizontally polarized and therefore reduce glare from sunlight reflecting off horizontal surfaces.

Photons

Photons may be described as packets of light energy, and scientists use this concept to refer to the particle-like aspect of light. Photons are unlike conventional particles, such as specks of dust or marbles, however, in that they are not limited to a specific volume in space or time. Photons are always associated with an electromagnetic wave of a definite frequency. In 1900 the German physicist Max Planck discovered that light energy is carried by photons. He found that the energy of a photon is equal to the frequency of its electromagnetic wave multiplied by a constant called *h*, or Planck's constant.

This constant is very small because one photon carries little energy. Using the watt-second, or joule, as the unit of energy, Planck's constant is 6.626 x 10^{-34} (a decimal point followed by 33 zeros and then the number 6626) joule-seconds in exponential notation. The energy consumed by a one-watt light bulb in one second, for example, is equivalent to two and a half million trillion photons of green light. Sunlight warms one square meter at the top of Earth's atmosphere at noon at the equator with the equivalent of about 14 100-watt light bulbs. Light waves from the Sun, therefore, produce a very large number of photons.

Sources of Light

Sources of light differ in how they provide energy to the charged particles, such as electrons, whose motion creates the light. If the energy comes from heat, then the source is called incandescent. If the energy comes from another source, such as chemical or electric energy, the source is called luminescent

Photoelectric Effect

The photoelectric effect is a process in which an atom absorbs a photon that has so much energy that the photon sets one of the atom's electrons free to move outside the atom. Part of the photon's energy goes toward releasing the electron from the atom. This energy is called the activation energy of the electron. The rest of the photon's energy is transferred to the released electron in the form of motion, or kinetic energy. Since the photon energy is proportional to frequency, the released electron, or photoelectron, moves faster when it has absorbed high-frequency light.

Metals with low activation energies are used to make photodetectors and photoelectric cells whose electrical properties change in the presence of light. Solar cells use the photoelectric effect to convert sunlight into electricity. Solar cells are used in place of electric batteries in remote applications like space satellites or roadside emergency telephones. Hand-held calculators and watches often use solar cells so that battery replacement is unnecessary.

Photochemical Detection

The change induced in photographic film exposed to light is an example of photochemical detection of photons. Light induces a chemical change in photosensitive chemicals on film. The film is then processed to convert the chemical change into a permanent image and to remove the photosensitive chemicals from the film so it will not continue to change when it is viewed in full light. Human vision works on a similar principle. Light of different frequencies causes different chemical changes in the eye. The chemical action generates nerve impulses that our brains interpret as colour, shape, and location of objects.

BEHAVIOUR OF LIGHT

Light behaviour can be divided into two categories: how light interacts with matter and how light travels, or propagates through space or through transparent materials. The propagation of light has much in common with the propagation of other kinds of waves, including sound waves and water waves.

Interaction with Material

When light strikes a material, it interacts with the atoms in the material, and the corresponding effects depend on the frequency of the light and the atomic structure of the material. In transparent materials, the electrons in the material oscillate, or vibrate, while the light is present. This oscillation momentarily takes energy away from the light and then puts it back again. The result is to slow down the light wave without leaving energy behind.

Denser materials generally slow the light more than less dense materials, but the effect also depends on the frequency or wavelength of the light. Under certain laboratory conditions, scientists can slow light down. In 2001 scientists brought a beam of light to a halt by temporarily trapping it within an extremely cold cloud of sodium atoms. Materials that are not completely transparent either absorb light or reflect it. In absorbing materials, such as dark colored cloth, the energy of the oscillating electrons does not go back to the light. The energy instead goes toward increasing the motion of the atoms, which causes the material to heat up.

The atoms in reflective materials, such as metals, re-radiate light that cancels out the original wave. Only the light re-radiated back out of the material is observed. All materials exhibit some degree of absorption, refraction, and reflection of light. The study of the behaviour of light in materials and how to use this behaviour to control light is called optics.

Refraction

Refraction is the bending of light when it passes from one kind of material into another. Because light travels at a different speed in different materials, it must change speeds at the boundary between two materials. If a beam of light hits this boundary at an angle, then light on the side of the beam that hits first will be forced to slow down or speed up before light on the other side hits the new material. This makes the beam bend, or refract, at the boundary. Light bouncing off an object underwater, for instance, travels first through the water and then through the air to reach an observer's eye. From certain angles an object that is partially submerged appears bent where it enters the water because light from the part underwater is being refracted.

The refractive index of a material is the ratio of the speed of light in a vacuum to the speed of light inside the material. Because light of different frequencies travels at different speeds in a material, the refractive index is different for different frequencies. This means that light of different colors is bent by different angles as it passes from one material into another. This effect produces the familiar colorful spectrum seen when sunlight passes through a glass prism. The angle of bending at a boundary between two transparent materials is related to the refractive indexes of the materials through Snell's Law, a mathematical formula that is used to design lenses and other optical devices to control light.

Reflection

Reflection also occurs when light hits the boundary between two materials. Some of the light hitting the boundary will be reflected into the first material. If light strikes the boundary at an angle, the light is reflected at the same angle, similar to the way balls bounce when they hit the floor. Light that is reflected from a flat boundary, such as the boundary between air and a smooth lake, will form a mirror image. Light reflected from a curved surface may be focused into a point, a line, or onto an area, depending on the curvature of the surface.

Scattering

Scattering occurs when the atoms of a transparent material are not smoothly distributed over distances greater than the length of a light wave, but are bunched up into lumps of molecules or particles. The sky is bright because molecules and particles in the air scatter sunlight. Light with higher frequencies and shorter wavelengths is scattered more than light with lower frequencies and longer wavelengths. The atmosphere scatters violet light the most, but human eyes do not see this colour, or frequency, well.

The eye responds well to blue, though, which is the next most scattered colour. Sunsets look red because when the Sun is at the horizon, sunlight has to travel through a longer distance of atmosphere to reach the eye. The thick layer of air, dust and haze scatters away much of the blue. The spectrum of light scattered from small impurities within materials carries important information about the impurities. Scientists measure light scattered by the atmospheres of other planets in the solar system to learn about the chemical composition of the atmospheres.

TRAVELS OF LIGHT

The first successful theory of light wave motion in three dimensions was proposed by Dutch scientist Christiaan

Huygens in 1678. Huygens suggested that light wave peaks form surfaces like the layers of an onion. In a vacuum, or a uniform material, the surfaces are spherical. These wave surfaces advance, or spread out, through space at the speed of light. Huygens also suggested that each point on a wave surface can act like a new source of smaller spherical waves, which may be called wavelets, that are in step with the wave at that point.

The envelope of all the wavelets is a wave surface. An envelope is a curve or surface that touches a whole family of other curves or surfaces like the wavelets. This construction explains how light seems to spread away from a pinhole rather than going in one straight line through the hole. The same effect blurs the edges of shadows. Huygens's principle, with minor modifications, accurately describes all forms of wave motion.

Interference

Interference in waves occurs when two waves overlap. If a peak of one wave is aligned with the peak of the second wave, then the two waves will produce a larger wave with a peak that is the sum of the two overlapping peaks. This is called constructive interference. If a peak of one wave is aligned with a trough of the other, then the waves will tend to cancel each other out and they will produce a smaller wave or no wave at all. This is called destructive interference.

In 1803 English scientist Thomas Young studied interference of light waves by letting light pass through a screen with two slits. In this configuration, the light from each slit spreads out according to Huygens's principle and eventually overlaps with light from the other slit. If a screen is set up in the region where the two waves overlap, a point on the screen will be light or dark depending on whether the two waves interfere constructively or destructively. If the difference between the distance from one slit to a point on the screen and the other slit to the same point on the screen is an exact number of wavelengths, then light waves arriving at that point will be in step and constructively interfere, making the point bright. If the difference is an exact odd number of half wavelengths, then light waves will arrive out of step, with one wave's trough arriving at the same time as another wave's peak. The waves will destructively interfere, making the point dark. The resulting pattern is a series of parallel bright and dark lines on the screen.

Instruments called interferometers use various arrangements of reflectors to produce two beams of light, which are allowed to interfere. These instruments can be used to measure tiny differences in distance or in the speed of light in one of the beams by observing the interference pattern produced by the two beams.

Holography is another application of interference. A hologram is made by splitting a light wave in two with a partially reflecting mirror. One part of the light wave travels through the mirror and is sent directly to a photographic plate. The other part of the wave is reflected first toward a subject, a face for example, and then toward the plate. The resulting photograph is a hologram. Instead of being an image of the face, it is an image of the interference pattern between the two beams.

A normal photograph records only the light and dark features of the face and ignores the positions of peaks and troughs of the light wave that form the interference pattern. Since the full light wave is restored when a hologram is illuminated, the viewer can see whatever the original wave contained, including the three-dimensional quality of the original face.

Diffraction

Diffraction is the spreading of light waves as they pass through a small opening or around a boundary. Young's principle of interference can be applied to Huygens's explanation of diffraction to explain fringe patterns in diffracted light. As a beam of light emerges from a slit in an illuminated screen, the light some distance away from the screen will consist of overlapping wavelets from different points of the light wave in the opening of the slit.

When the light strikes a spot on a display screen across from the slit, these points are at different distances from the spot, so their wavelets can interfere and lead to a pattern of light and dark regions. The pattern produced by light from a single slit will not be as pronounced as a pattern from two slits. This is because there are an infinite number of interfering waves, one from each point emerging from the slit, and their interference patterns overlap one another.

Chapter 3

Electric Power System

The Earth is a huge magnetic sphere. Living things have been tied to the Earth's natural magnetic field from the time that life began. This magnetic field permeates and contributes to all life on the planet by the generation of what we call the atmosphere. This field varies in strength and consistency through the ages and, with this variance, so too does life on the planet change. Our bodies own magnetic frequencies and bio-field patterns react to this variance of the Earth's field.

All matter on Earth assists in creating this field and so becomes charged with this magnetic resonance, displaying different results, dependant on the substance. Each individual cell—as well as entire organisms senses and derives timing information from the natural cycles of the geomagnetic field. As all matter is made up of magnetic resonant field patterns, of varying strength and frequencies, applying a magnetic device of any configuration will produce an effect one way or another.

The healers-of-old knew the benefits and curative powers fo magnetic *lodestones* when properly placed along the meridian lines of energy and not necessarily at the area of pain. Lodestones, being basically *iron oxide,* have the capacity to retain and exhibit a magnetic field. These lodestones have been charged by the Earth, and thus are harmoniously in tune with the human bodies' frequencies as well, they have for thousands of years been successfully used in the healing arts. These healers continued their practice through the ages, since the patients could see the results and then utilized the methodology themselves. Although this same methodology is still practiced with desired results in some cultures today, twenty-first century man is destroying the healing effects with modern materials.

Modern magnetic therapy is based on traditional ideas, but uses stronger and "out of phase" magnetic materials. Most everyone considers magnetism a basic and natural force similar to the magnetic field exhibited by the field of an artificial permanent or electromagnet has a distinct man-made flavour within the internal structure or matrix of the field itself. The frequency of the field is the governing factor as to whether the field is beneficial or harmful.

The field of a permanent magnet receives its frequency when energized, depending on the energizing process and the composition of the magnet itself, resulting in the frequency of the final product. Powerful permanent magnets have their matrix altered by the artificial energizing processes used, and do not conform to the natural complex magnetic field structure of the human bio-organism, or that of the planet. These processes use artificial electrical energy, which has within its structure a frequency composition different to that of nature. This is why lodestones work well in healing situations since they have the required field strength and attenuation to the life force of the planet.

Human use of electromagnetism for power and communications has produced an abnormal electromagnetic environment unlike anything that existed before. More than a thousand scientific papers on this topic are published each year. Both epidemiological and laboratory studies exist, indicating that abnormal frequencies, regardless of their frequencies, produce increased health risks.

AN ELECTROMAGNETIC FIELD

All electromagnetic fields are force fields, carrying energy and capable of producing an action at a distance. These fields have characteristics of both *waves* and *particles*. An electric current flowing in a wire or coil produces its own magnetic field. A compass needle will swing at right angles to the wire. What is the difference between an *electric* field and a *magnetic* field? Actually both are usually all around us, since they are present wherever there is electricity. Both an electric and a magnetic field exist around power lines, appliances, light fixtures, and electric wiring; but electromagnetic fields are generated only when current is flowing through a wire. The voltage on a wire produces electric fields.

If you plug in a lamp or an appliance, even though it is not turned on, it will produce an electric field. When the light is lit or the appliance is running, both an electric and electromagnetic field will be present. A permanent magnetic field is stationary or still, unless you are moving the magnet. Permanent magnets are completely safe, and are not involved in this controversy.

Magnetic fields are produced by moving electrical currents, and the field of a bar magnet is produced by the spinning electrons around the atomic nuclei in the iron material of the magnet itself. In order to act as a "magnet," these atoms must be lined up in the same direction so that the individual magnetic fields combine to produce one big field. If a magnet is dropped on a hard floor, the magnetic field is destroyed because the atoms are "shaken up" and their direction is changed, resulting in a random pattern.

When an electrical current flows in a wire, the movement of the electrons produces a similar field in space, but one that is oriented around the wire. If the current is a DC (direct current), the magnetic field is steady, like that form a permanent magnet. The strength of the magnetic field depends upon the amount of current flowing in the wire—the more current there is, the stronger the magnetic field will be.

If the current in the wire is fluctuating, the magnetic field will have the same fluctuations. We characterize the field in this case on the basis of the rate, or frequency of fluctuation. All of the possible frequencies of electromagnetic waves or fields can be put onto an electromagnetic spectrum, starting with the slowest frequencies and going to the highest. The electromagnetic spectrum is usually pictured as a line that increases from zero, to trillions of cycles per second. About three-fourths up the frequency spectrum is something we are very familiar with—light. Sunlight, as well as the light from electric lights, is the same thing as the magnetic field. It is produced by the movement of electrons and has the same characteristics as the Earth's magnetic field, radio waves, and X-rays. It is an electromagnetic field, but one for which we have developed specific anatomical detectors—our eyes. Electrical currents flow in living organisms and they produce magnetic fields that extend outside of the body and can be influenced by external magnetic fields as well. The magnetic field produced by the brain is easily detected using a special magnetometer.

Extra low frequency magnetic fields, such as those from 50 & 60 Hz systems, penetrate even a cast iron bath tub with little attenuation. They freely intrude our homes and our bodies. They are biologically active at intensities of a milligauss (mG) or less.

The flow of the electrical currents in the brain produce a magnetic field that can be measured and analyzed several feet away from the head. We, and all other living things, are surrounded by a magnetic field extending out into space from our bodies, and the fields from the brain reflect what is happening in the brain. Living organisms have the ability to somehow sense these minute daily cycles in the Earth's magnetic field and to use them to time their biological cycles.

ELECTRONICS

If two electrons are separated by one centimeter, each electron will experience a repulsive force of 2.3×10^{-24} gram equivalents. Doubling the separation will reduce the force to one quarter. Doubling the number of electrons on one side will double the repulsive force acting on the lone electron on the other side. Most matter is *neutral,* so it contains an equal number of protons and electrons. However, sometimes electrons are deliberately or accidentally removed from one object, and deposited onto another object.

If a proton is placed near these two charged objects, the proton will be attracted to the negative object, which has excess electrons, and simultaneously repelled from the positive one, which is missing some electrons. The total force acting on the proton is a measure of the electric field that the proton is exposed to. The direction of the force that acts on the proton is the same as the direction of the electric field and the strength of the force is proportional to the strength of the electric field. If an *electron* were substituted for the *proton*, the force would be of the same strength, but in the opposite direction of the electric field. If two protons "tied" together were substituted for the single proton, they would experience twice as much total force.

If the terminals of a battery are connected to two large metal plates, the + plate will lose some electrons, which will flow through the battery to the–plate. A battery acts as an electron pump, establishing an electric field between the plates. A proton would be *attracted* toward the-plate and *repelled* from the + plate. The strength of this electric field equals the battery voltage divided by the distance between the plates (usually measured in meters). For example, if the battery is 1.5 volts and the spacing is one meter, the electric field is 1.5 volts per meter (V/m), but if the spacing is reduced to ½ meter, the electric field increases to 3V/m, which doubles the force that a single proton, if placed in the centre, would experience. Any charged particle placed between the plates would experience a force *downward* if the particle were positive, or *upward* if negative.

All conductive bodies contain charged particles that can freely move around. In metals these particles are the *electrons* from the outer part of the atom. In conductive liquids such as salt water, the conductive particles are called *ions*: positively charged *sodium* atoms and negatively charged *chlorine* atoms, which can move through the water. If a conductive body is placed between the plates, the charged particles inside the conductor will quickly shift position a little. The charged particles shift position simply because they are responding to the electric field from the battery and metal plates.

Because no electric field now exists within the conductive body, the internal charged particles have stopped moving and the configuration is stable. If the battery is subsequently reconnected to the plates in reverse, all charges will reorient so that–and + ultimately interchange after a small fraction of a second. The movement of charges is defined as an *electric current*, so that during that brief time the charges in the conductive body are reorienting themselves, a current is flowing. At all other times (when charges are stable), no current flows. Therefore, inside a conductive body, no current flows and no internal electric field exists except during the periods when the outside electric field is changing.

A changing *electric* field can produce a *magnetic* field. The electric field is fundamental; the magnetic force arises directly from the electric force. In two parallel wires each 40" (100 cm) long and are one cm apart and carry one amp of current will attract each other with a force of 0.002-gram equivalents due to the magnetic force. The force will double if the current in either wire is doubled or if the separation is reduced to ½ cm. If the direction of current in one of the wires is reversed, the force will become repulsive, but will remain the same strength.

The current in the first wire creates a magnetic field that acts on the current in the second wire. In a similar way, the second wire also acts on the first. The attraction is mutual; the first wire attracts the second with the same amount of force that the second attracts the first. No charge will actually flow through the air gap between them, yet a magnetic field will be created in that gap as if the charge were flowing. The amount of magnetic force a wire experiences is proportional to the product of magnetic field strength that the wire is subjected to, times the length of the wire, times the amount of current (in amps) the wire is carrying. The direction of the force is at right angles both to the direction of the magnetic field and the direction of the current.

Besides being influenced by a magnetic field, a wire that is carrying a current also produces a magnetic field. If the wire is long, the field strength is double at half the distance from the wire and is four times as much at a quarter the distance. *Milligauss* is a unit of *flux density*, not magnetic field strength, but the difference is not important. A wire that carries one amp will have a magnetic field of 200 milligauss (mG), a distance of one cm away. At two cm the field will be one-half as much. Certain types of metaliron, steel, nickel, and vanadium internally produce this circulating current without the help of a battery or wires.

These metals are called *ferromagnetic* and their single outer electrons each produce a tiny, circulating current due to the electron orbits. All electrons in a Ferro magnet (if fully magnetized) spin in the same direction. If two such magnets are brought close together with the *North* of one facing the *South* of the other, they will *attract*. This attraction is because the electron currents spin in the same direction around the edge of the magnets. If the North of a magnet is brought close to the North of another magnet, they will *repel*, because the electron currents are circulating in opposite directions.

A changing *magnetic* field can create an *electric* field. This is the basis of power transformers and electric power generation. One coil of wire is placed close to another. An increasing current flow is established in the first coil, which creates an increasing magnetic field. The increasing magnetic field causes electrons to circulate (current) around the second coil in such a way as to oppose the increasing magnetic field that the first coil is trying to establish.

The changing magnetic field creates an electric field directly in the wire, which causes electrons in the wire to move. The amount of voltage generated is proportional to the total length of wire in all the loops. A changing magnetic field will also create an electric field in the human body, which causes a current to flow. The current is at right angles to the direction of the changing magnetic field, and is strongest near the extremities of a person. Current tends to flow along the outer surface of the body. Its strength is proportional to the rate at which the field changes.

For a 60 Hz (60 oscillations per second) field with strength of three milligauss, a current is created in an adult, which is on average, one-billionth amp per square centimeter of cross-sectional area. For these currents induced by *magnetic* fields, the current per area increases proportional to the linear size of the organism. An animal only 1/10 the size of an adult human would average 1/10 as much current per area. An AC *electric* field will also create a current in the body. This current usually flows in the same direction as the electric field and is fairly uniform throughout the body. As with the magnetic field current, the current from an AC electric field is proportional to the frequency multiplied by the strength. For an adult, a 2000 V/m electric field at 60 Hz will create, on average, one-billionth amp per square cm. Unlike the case of magnetic current, the electric current per square centimeter does not scale down for a small animal.

The changing electric and magnetic effects make possible electromagnetic waves (radio waves). These waves are usually emitted from a wire (antenna) connected to an oscillator (in essence, a battery changing polarity very fast). The antenna rapidly oscillates between + and –, creating a rapidly changing electric field. Most radio waves are in the frequency range from half a million to several billion oscillations per second. That rapidly changing *electric* field creates a rapidly changing *magnetic* field.

The magnetic field, in turn, creates an electric field, which reinforces the existing electric field. Because of this reinforcement, the strength of the electric field far from the antenna is much greater than it would be without the reinforcement. For example, if a one-foot antenna is connected to one volt DC (electric field is not oscillating), the electric field one mile away will be 1/100 billion V/m.

However, if that same one volt is continuously reversed 200 million times per second (proper frequency for a one foot antenna), the peak electric field one mile away will be 1/3000 V/m, or 33 million times higher. By using radio waves, the fields can be carried a long distance. For the AC frequencies typically encountered in the home (60, 180, and 300 oscillations per second are typical from power line current, and up to 17,000-35,000 oscillations from the horizontal scan of TV and computer monitors), antennas and coils would have to be at least several miles long to emit true radio waves.

Because the devices that emit these fields (vacuum cleaners, fluorescent lights, etc.) are much smaller than that, household sources of fields are often pure magnetic (from wires or coils that carry AC current), pure electric (from metal plates or wires that are connected to the electric "hot" wire, regardless of whether any current is present), or some mix of magnetic and electric.

NATURAL ELECTROMAGNETIC FIELDS

The spinning core of molten iron miles beneath the surface of the Earth creates a *dipole* magnetic field, like a bar magnet. The energy of the sun distorts and perturbs this simple field into a unique structure known as the *magnetosphere*. The strength of the Earth's magnetic field averages about half a gauss. Its daily change in strength is less than 0.1 gauss. Compared with the 200-gauss strength of the magnet that holds a refrigerator door closed, this seems miniscule. The sun constantly gives off a solar wind, composed of high-energy atomic particles.

These particles travel through space at great speed and crash into the outer layers of the Earth's magnetic field, compressing it until its energy matches that of the solar wind. The area of interaction between these two forces is called the *bow-shock region*. The sun's magnetic field is organized into sectors, much like the segments of an orange. Alternate sectors have their fields directed inward and outward, so that as the sun rotates, a slight change is produced as each sector boundary crosses the line connecting the sun with the Earth.

On the side away from the sun, the magnetic field is drawn out into a long *magneto tail*, which stretches away from the Earth far into space. It contains vast stores of energy and demonstrates reliable daily fluctuations in strength, as well as longer periodic changes. It is subject to sudden, massive storms produced by energetic events on the sun. Over geologic time, it exhibits strange reversals of polarity.

The Earth rotates on a day-night cycle within this complex field. The magnetosphere does not rotate but remains fixed in space, with one side always facing the sun. Because of this, any given spot on the surface of the Earth is in a constantly changing magnetic field. The daily rise and fall in the strength of this field causes biological rhythms. Magnetic storms appear to have a direct effect on the operations of the human brain. This interaction results in the generation of enormous electrical currents with billions of watts. It also results in the production of ionizing radiation and various electromagnetic waves. The EMF energy of the earth is usually absorbed through the feet. There is a point on the sole near the upper centre of the foot, through which it is specifically drawn up. Lying on the ground is also a way of recharging. The earthly forces rebalance the heavenly forces.

Water can be energized with specific magnetic fields that aid life compared to untreated water. Water that comes up out of the earth, or is brought up out of the earth, looses it's charge after ninety-two hours and thus any water having magnetic healing properties, such as the water from Lourdes, must be used soon after taking the sample from its source. Any water brought up from an underground source and used to energize a treatment device, looses all effectiveness in ninety-two hours as well.

If the treatment device also contains magnets or a mechanism to spiral the water (clockwise in Northern hemisphere), then a small helpful effect is imparted to the water for a limited time. Magnetic field generators are available that impart a magnetic field to the water, designed purposefully for bio-magnetic field patterns based on the frequencies of the human bio-field. This water is then used in the treatment of ailments, with positive results. With life on Earth being water-based, magnetizing water before use, whether on plants or humans, will give better health and vitality.

THE MAGNETIC ORGAN

In 1975, a researcher noticed that the bacteria he was studying always clustered at the north side of their culture dish. Even if he turned the dish so that they were at the south end and left it overnight, the next morning the bacteria were back at the north side. Each bacterium contained a chain of tiny magnets! The magnets were actually crystals of the natural magnetic mineral *magnetite*, the original lodestone of preliterate peoples. Somehow, the bacteria absorbed the soluble components from the water and put them together in their bodies as the insoluble crystalline chain.

In 1971, another researcher reported on a lengthy series of experiments with pigeons. He found the same crystals of magnetite, as a submicroscopic mass located on the surface of the pigeon's brain. He found that the mass of crystals was full of nerve fibers that seemed to go into the brain. The magnetite crystals "tell" the pigeon's brain the exact direction of the Earth's magnetic field, and the pigeon uses this information to navigate with its amazing precision. The salamander has two separate magnetic navigational systems. One provides a simple compass, so that when traveling "cross country," it will go in a straight line.

The other system enables it to return to the exact spot where it was hatched from its egg in order to mate and lay more eggs. The human "magnetic organ" has been found in the posterior wall of the ethmoid sinus, just in front of the pituitary gland. Humans have an innate ability to sense the direction of magnetic north, and this ability can be blocked by placing a bar magnet against a person's forehead for only fifteen minutes! The directional sense is disturbed for as long as two hours after application of the magnet. It is a sense organ that informs the organism of the direction of the Earth's magnetic field. Nature has provided us with yet another organ that also senses the field and extracts even more significant information from it.

THE PINEAL GLAND'S MAGNETIC SENSE

The *pineal gland* is the "clock" that controls biological cycles. The cyclic pattern of sleep-wakefulness is dependent upon the level of *melatonin* secretion by the pineal. The pineal gland is the principal structure in the brain that is directly sensitive to the Earth's magnetic field. As a result, it functions abnormally when exposed to abnormal fields. Because the pineal produces a host of psychoactive chemicals (such as *melatonin, dopamine, serotonin,* and others), its abnormal functioning may result in a variety of neurological and behavioral abnormalities. It was first determined that a part of the output of the retina is diverted to the pineal, where it senses the day-night cycle and melatonin secretion is adjusted accordingly.

The pineal is also sensitive to the daily cyclic pattern in the Earth's magnetic field. Abnormal secretion of hormones by the pineal is linked to many behavioral abnormalities. The primary effect of a chronically abnormal biocycle is the production of *chronic stress syndrome* a condition that produces a wide variety of health problems, including a marked decline in competency of the immune system. Nature intended the pineal to simultaneously receive the same signals from the daily pattern of day-night and the same rise and fall in strength of the geomagnetic field.

Obviously, when one or both signals are abnormal, the pineal does not respond in the normal fashion, and the body's biological cycles become disturbed—with important health results. Over two billion years of evolution, living things have taken advantage of the two portions of the electromagnetic spectrum that could be depended upon to always be present: the geomagnetic field and visible light. Life developed specific organs to sense the geomagnetic field and to derive timing information from it just as it developed specific organs to sense and derive information from light.

THE HUMAN CRYSTAL

The key to understanding the assimilation of energy into our physical structure is through the awareness of our bodies as a series of *synchronous, interacting crystal structures*. The body on this level is linked by oscillating solid and liquid crystals that form an overall energy pattern for the total body. Each organ, gland, nerve system, cell and protein structure—even the tissue salts in the body—shows a level of organization with some degree of crystalline-like function. Marcel Vogel, the world-renowned crystal expert, has pointed out that the human energy field exists as an array of oscillating energy points which have a layered structure and a definite symmetry.

He points out that these properties fulfill the definition of a normal crystal in material form. Our bone structure has long been recognized as a solid crystal structure with *piezoelectric* properties. A piezoelectric effect is the creation of an electromagnetic field pulse when a crystalline structure is physically stressed or pushed out of its normal shape. As a solid crystal, it has the ability to convert vibrational energy, such as sound or light, into magnetic and electric energy.

Crystals can absorb, store, convert, amplify, transduce and transmit vibrational energies, which have biological effects. Psychic healing energies alter the measurable electromagnetic pattern of a crystal, suggesting that subtle energies can alter the physical structure of a crystal. When physical stress or an electromagnetic field (EMF) is applied to a piezoelectric crystal, the crystal will change shape and generate an EMF. Bone, quartz crystal and tourmaline are among the few crystal forms with piezoelectric properties.

Studies suggest that the crystalline-like components of the extra cellular matrix of bone, such as *collagen* and *proteoglycans*, possess piezoelectric qualities. Mineralized tissue such as cartilage, dentin, teeth in general and relatively non-mineralized tissues such as *keratin* in skin, *elastin*, artery tissue, connective tissue (tendons and ligaments), and even some amino acid crystals (*glysine, proline* and *hydroxyproline*) all have piezoelectric properties.

The main·forces which create pulsed piezoelectric EMFs in bone are the anti-gravity muscles, the cardiovascular system, voluntary muscles and impact with the environment. Projected EMFs have the ability to create a piezoelectric response in bone. These fields have biological activity affecting: cell nutrition, local pH control, enzyme activation and suppression, orientation of intra and extra cellular macromolecules, migratory and reproductive activity of cells, synthetic capacity and specialized function of cells, contractility and permeability of cell membranes, and energy transfer.

Another type of electric or electromagnetic field response is also found in bone structure, called *streaming potentials*, an electrical field created by the flow of ions, charged solutes and cells such as red blood cells through the tissues, carried by extra cellular fluids such as blood through the extra cellular matrix. When the bone is even subtly bent from pressure by such events as walking or even by the pulse of our arteries, the extra cellular fluids are pumped through the bone. A charge is created by the electrostatic interaction of the charged fluids moving past the fixed charge in the crystalline bone structure. These fluids interact with the piezo fields of bone.

LIQUID CRYSTALS

There are other forms of crystalline-like systems in the body, and one of the most important is the *liquid crystal*. A liquid crystal is technically defined as having form, liquid properties, stored information and a measurable electromagnetic field. A liquid crystal can act simultaneously as *liquid* and *crystal*. The larger liquid crystal systems include fatty tissue, muscle and nerve tissue, the lymphatic system, white blood cells, and the various pleural and peritoneal linings.

Muscle and nerve tissue exist as liquid crystal systems held in shape by bone and skin systems. The muscles have also been shown to have some piezoelectric properties. On a cell level, all cells and cell membranes are considered liquid crystals. These include the plasma membrane, mitochondrial membrane, smooth and rough endoplasmic reticulum, nuclear membranes and chloroplast membranes. Bodily fluids also have crystal qualities. A water molecule contains the potential forms of all crystals in its primary form of a *tetrahedron*. Water can bring all different forms of ions into a crystalline state an hold them in solution.

The more structured the water is, the higher concentration of ions it can hold. One of the most important ion solutions is the dissolved cell salts. When the body cells and tissues become diseased or cancerous, the crystalline protein structures no longer have the proper configuration to maintain the water in a structured state.

Ions and other solutes in the cell consequently become redistributed by the new pattern of unstructured water. If the water in the cell, extra cellular fluid or blood plasma becomes structured, it will then be able to attract and hold more ions by virtue of its hydration shell patterns. Identical crystal resonant fields emanated by micronutrients are attracted to larger resonant fields emanated by the organism's larger crystalline and liquid crystalline patterns. This explains how micronutrients, homeopathic remedies, etc., through vibrating crystalline fields, are drawn to the appropriate resonating crystalline sites. In *An Atlas of Cellular Oscillators* by P.R. Rapp, research in over 450 papers is cited in cataloging an atlas of biological and biochemical oscillators with a periodicity of one hour or less.

This includes oscillations in enzyme-catalyzed reactions such as photosynthesis, oscillations in protein synthesis, and oscillations in cell membranes, secretory cells, neuronal cells, skeletal cells, smooth cells, heart muscle cells and cell movement. In a healthy state, the body's structures are a multi-leveled series of interacting systems and subsystems that resonate harmoniously. From this perspective, disease occurs when this synchronicity is thrown into disharmony. Each organ system or subsystem gives off a specific measurable electromagnetic field. The EMFs are measurable, subtle fields that can have great effect on the behaviour of an organism.

MAN-MADE ELECTROMAGNETIC FIELDS

Life on earth evolved under geomagnetic fields and electromagnetic forces that have been constant for hundreds of millions of years. All that changed a little over 100 years ago, when Thomas Edison built the first power station in New York City. Before 1900, the Earth's electromagnetic field was composed simply of the field and its associated micro pulsations, visible light, and random discharges of lightning. We swim in a sea of energy that is almost totally man-made. Silicon sensors are already calling to each other. Soon, countless communicating microchips, embedded everywhere, will be sending streams of encoded electrical energy through glass, steel, concrete, bone and flesh.

The exposure of living organisms to abnormal

electromagnetic fields results in significant abnormalities in physiology and function. The cells of the brain and central nervous system are particularly sensitive to abnormal fields. The developing central nervous system of the fetus or the newborn is particularly sensitive. These abnormal fields arise primarily from external sources. Most are unaware of the substantial risks associated with the use of many devices in their own homes. 60-Hz fields of only 3 milligauss have been shown to be significantly related to increase in cancer rates.

Microwaves

Microwaves are very short waves of electromagnetic energy that travel at the speed of light. Microwave ovens contain a *magnetron,* a tube in which electrons are affected by magnetic and electric fields in such a way as to produce micro wavelength radiation at about 2450 Megahertz (MHz) or 2.45 Gigahertz (GHz). This microwave radiation interacts with the molecules in food. All wave energy changes polarity from positive to negative with each cycle of the wave. In microwaves, these polarity changes happen millions of times a second. Food molecules—especially the molecules of water—have a positive and negative end in the same way a magnet has a north and a south polarity.

As the microwaves generated from the magnetron bombard the food, they cause the polar molecules to rotate at the same frequency, millions of times a second. All this agitation creates molecular friction, which heats up the food. The friction also causes substantial damage to the surrounding molecules, often tearing them apart or forcefully deforming them.

The scientific name for this deformation is *structural isomerism*. By comparison, microwaves from the sun are based on principles of pulsed *direct current* (DC) that don't create frictional heat; microwave ovens use *alternating current* (AC) creating frictional heat. A microwave oven produces a spiked wavelength of energy with all the power going into only one narrow frequency of the energy spectrum. In addition to violent frictional heat effects (called *thermic effects*), there are also *athermic effects*, which have hardly ever been taken into account. These athermic effects also deform the structures of molecules and have qualitative consequences. For example, the weakening of cell membranes by microwaves is used in the field of gene-altering technology.

Because of the force involved, the cells are actually broken, thereby neutralizing the electrical potentials—the very life of the cells—between the outer and inner sides of cell membranes. Impaired cells become easy prey for microorganisms. Natural repair mechanisms are suppressed, and cells are forced to adapt to a state of energy emergency they switch from *aerobic* to *anaerobic* respiration. Instead of *water* and *carbon dioxide*, *hydrogen peroxide* and *carbon monoxide* are produced. Any reversal of healthy cell processes may occur because of a number of reasons, and our cells then revert from a robust *oxidation* to an unhealthy *fermentation*.

The same violent friction and athermic deformations, that can occur in our bodies when we are subjected to radar or microwaves, happens to the molecules in the food cooked in a microwave oven.

Microwave Sickness

The Russians did research on thousands of workers who had been exposed to microwaves during the development of radar in the 1950's. Their research showed a breakdown of the human life-energy field, so serious that the Russians set strict limits of 10 microwatts exposure for workers and one microwatt for civilians. The first signs of microwave sickness are low blood pressure and slow pulse. The later and most common manifestations are chronic excitation of the sympathetic nervous system (stress syndrome) and high blood pressure.

This phase also often includes headache, dizziness, eye pain, sleeplessness, irritability, anxiety, stomach pain, nervous tension, inability to concentrate, hair loss, plus an increased incidence of appendicitis, cataracts, reproductive problems, and cancer. The chronic symptoms are eventually succeeded by crisis of adrenal exhaustion and ischemic heart disease.

Changes are observed in the blood chemistries and the rates of certain diseases among consumers of microwaved foods. Lymphatic disorders were observed, leading to decreased ability to prevent certain types of cancers.

An increased rate of cancer cell formation was observed in the blood. Increased rates of stomach and intestinal cancers were observed. Higher rates of digestive disorders and gradual breakdown of the systems of elimination were observed. There is degeneration and destabilization of the external energy activated potentials of food utilization within the processes of human metabolism and of internal cellular membrane potentials while transferring catabolic processes into the blood serum from the digestive process.

There is a degeneration and breakdown of nerve electrical circuits and loss of energy field symmetry in the neuroplexuses both in the front and the rear of the central and autonomic nervous systems, along with a loss of balance and circuiting of the bioelectric strengths within the ascending *reticular activating system* (controls the function of consciousness). There is long-term cumulative loss of vital energies within humans, animals and plants that are located within a 500-meter radius of operational equipment. There is long-lasting residual effects of magnetic "deposits" located throughout the nervous system and lymphatic system.

There is destabilization and interruption in the production and maintence of hormones in males and females by continually eating microwaved foods. Continually eating food processed from a microwave oven, or using cell phones, causes permanent brain damage by shorting out electrical impulses in the brain (depolarizing or demagnetizing the brain tissue). There is markedly higher levels of brainwave disturbance in the *alpha, theta,* and *delta* wave signal patterns of persons exposed to microwave emission fields, with a resultant negative psychological effects, including loss of memory, loss of ability to concentrate, suppressed emotional threshold, deceleration of intellective processes, and interruptive sleep episodes.

Microwaves cause long-term depolarization of tissue

neuroelectric circuits with irreversible damage to the neuroelectrical integrity of various components of the nervous system, in individuals subjected to continual range emissive field effects of microwave apparatus, either in cooking apparatus, in transmission stations, and PCS cell phones. The body cannot metabolize unknown byproducts created in microwaved food.

ELECTRIC POWER SYSTEMS

Our electric-power systems operate at fifty or sixty cycles per second, just above the highest naturally occurring frequency of 30 Hertz (Hz.)the frequency in cycles per second. Our microwave beams operate at billions of times per second and are getting ever closer to the trillion-cycle frequencies of visible light. We have filled the previously empty electromagnetic spectrum between these two extremes with man-made radiation that never existed on Earth, and we did it in less than 80 years. Because our society runs on electromagnetic (EM) energy, there really is no place to hide. We are living in a "radio-frequency soup."

Microwave and low-frequency radiation is all around us—emitted by television and radio towers, satellites, high-voltage electric lines, personal computers, police radios, paging systems, microwave ovens and electronic games, CB radios, alarm systems, telecommunications systems, garage door controls, signal generators, etc. Even in the remotest mountain valleys, you're exposed to some level of the ubiquitous power frequencies of 50 or 60 Hz., short-wave radio waves reflected back from the ionosphere, as well as microwaves being beamed from satellites. It is not ionizing—it doesn't displace electrons and start biochemical reactions in our cells like ionizing radiation—but it does affect human health.

Televisions and Computers

There have been "clusters" of miscarriages among female computer operators. Most people know that the television or computer screen gives off a small amount of ionizing radiation as X-rays, but few are aware of the amount and distribution of the electromagnetic radiation from the entire unit. Actually, the electromagnetic fields are generated within the circuitry of the entire device. The TV set is what is called a *broadband radiating source*, meaning that it gives off a variety of frequencies, from the 60-Hz power frequency to radio frequencies in the MHz (mega hertz) range.

The radiation is given off in all directions from the TV set and computer screen. The fly-back circuit operates in the VLF (very low frequency) range, generally around 17 kHz, and this frequency is a major part of the total frequencies given off. When you sit in front of the set, you are being irradiated with a wide range of frequencies, with the strongest components probably being the 60-Hz power frequency and the 17-kHz fly-back frequency. In general, the larger the size of the screen, the stronger the fields, and the further out from the unit they will extend.

Electromagnetic radiation goes through wood and other usual building materials. If the TV set or computer is placed with its back against an inside wall, radiation will be present in the adjoining room just as if there were no wall present. Consequently, an infant's or child's bed should not be placed against a wall opposite a television set, or computer, regardless of the device's field strength.

If you live in an apartment or condominium, you probably don't know what's on the opposite side of the walls or above or below you. Because the strength of the field drops off with the square of the distance, the closer one's head to the machine, the stronger the field.

A TV viewer usually sits back at least 6-8 feet from the screen. But with computers, the keyboards are attached to the rest of the device and the operator's head is only about fifteen to eighteen inches from the screen. There is a trend of introducing school children to computers at earlier and earlier grade levels. Many schools are using computers and people who have upgraded their own machines and given their older, unshielded, high-radiation models to a local school donated many to them.

Schoolchildren are still growing and have a constant level

of cell multiplication within their bodies, which leaves them even more susceptible to the harmful effects of electromagnetic field bombardment.

Fluorescent Lights

The incandescent lamp simply causes a resistive filament to glow when current at 120 volts and very small amperage is forced through it. The fluorescent lamp has no filament. Instead, a coating of chemical on the interior of the tube is excited to glow when a high-voltage discharge is produced within the tube. This requires that the 120 volts of house current be raised to several thousand volts by means of a transformer. The fluorescent light produces much different magnetic fields compared with the incandescent bulb. Two inches from a 60-watt incandescent bulb, the 60-Hz field is 0.3 milligauss; and at one foot, the field is lost in the ambient field.

A 10-watt fluorescent tube produces a magnetic field at least twenty times greater than a 60-watt incandescent bulb. The ceiling fluorescent fixtures, with several 20-watt tubes, produce a field greater than 1 milligauss near the heads of the office workers below. Considering the average time of exposure, the fields produced by fluorescent lights can reach dangerous levels.

Electric Clocks

An electric motorized clock, plugged into a wall socket, produces an amazingly high magnetic field because of the small electric motor that runs it. A small bedside alarm clock of this type will produce a field of as much as 5 to 10 milligauss two feet away. If the bedside table is placed close to the sleeper's head, or if a clock or radio is placed on the headboard shelf near the head, it is within this range. The dose rate is considerable for the average eight hours per night.

Razors and Hairdryers

The electric razor and hair dryer produce an extremely high-strength magnetic field (200 to 400 milligauss for the razor and 50 milligauss at six inches for the hair dryer) if it is a line-operated device—that is, if it is plugged into a wall socket rather than battery operated. These devices are used for just a few minutes once a day, so the total exposure, or dose, to the user is minimal.

Electric Blankets, Heating Pads and Water Beds

Electric-blanket and heating pad field strengths are somewhat lower (50-100 milligauss) but are still within the danger zone. The electric blanket or heating pad is applied as close to the body surface as the electric razor. Even the water bed heater, being farther away from the body, because it is used for several hours at a time, the total administered dose is much higher. The incidence of miscarriage is much higher among pregnant women who use electric blankets or heating pads than among those who do not.

Electric Heaters

A four-foot-long baseboard heater produces 23 milligauss at a distance of six inches; at one foot, the field is 8 milligauss; at two feet, it is about 3 milligauss; and it falls to 1 milligauss at three feet. The small portable electric heaters that plug into the wall socket produce about the same field intensities, but they are even more dangerous because they may be placed much closer to the body. In recent years, electric heat cables have been installed in ceilings as a more efficient means of heating an entire room. This produces an average field of about 10 milligauss throughout the entire room. Any device that electrically produces heat will have a major field strength at short distances from the device. The electric stove produces 50 milligauss at eighteen inches above a twelve-inch burner.

Microwave Ovens

Today, more than 90% of American homes have a microwave oven, used for meal preparation. The Nazis originally developed microwave radiomissor cooking ovens for use in their mobile support operations, to be used for the invasion of Russia. Microwaves are very short waves of electromagnetic energy that travel at the speed of light.

Microwave ovens contain a magnetron, a tube in which electrons are affected by magnetic and electric fields in such a way as to produce micro-wavelength radiation at about 2450 Megahertz (MHz) or 2.45 Gigahertz (GHz). This microwave radiation interacts with the molecules in food.

All wave energy changes polarity from positive to negative with each cycle of the wave. In microwaves, these polarity changes happen millions of times every second. Food molecules especially the molecules of water—have a positive and negative end in the same way a magnet has a north and a south polarity. As the microwaves generated from the magnetron bombard the food, they cause the polar molecules to rotate at the same frequency, millions of times a second. All this agitation creates molecular friction, which heats up the food.

The friction also causes substantial damage to the surrounding molecules, often tearing them apart or forcefully deforming them. The scientific name for this deformation is *structural isomerism*. By comparison, microwaves from the sun are based on principles of pulsed direct current (DC) that don't create frictional heat; microwave ovens use alternating current (AC) creating frictional heat. A microwave oven produces a spiked wavelength of energy with all the power going into only one narrow frequency of the energy spectrum.

The toxic effects of microwave cooking are another in a long list of unnatural additives in our daily diets. Artificially produced microwaves are based on the principle of alternating current. Atoms, molecules and cells hit by this hard electromagnetic radiation are forced to reverse polarity 1 to 100 billion times a second. Production of unnatural molecules is inevitable. Naturally occurring amino acids undergo isomeric changes (changes in shape morphing), as well as transformation into toxic forms, under the impact of microwaves produced in ovens.

There are no atoms, molecules or cells of any organic system able to withstand such a violent, destructive power for any extended period of time, not even in the low energy range of milliwatts. Current regulations allow leakage of 1 milliwatt per centimeter squared. Damage to the gasket material around the door will markedly increase the level of microwave radiation that leaks from the device when it is in use. Of all the natural substances, which are polar, the oxygen of water molecules reacts most sensitively. This is contrary to the conventional heating of food, in which heat transfers convectionally from without to within. Cooking by microwaves begins within the cells and molecules where water is present and where the energy is transformed into frictional heat.

EFFECTS ON CELLS

In addition to violent frictional heat effects (called *thermic* effects), there are also *athermic* effects, which have hardly ever been taken into account. These athermic effects also deform the structures of molecules and have qualitative consequences. For example, the weakening of cell membranes by microwaves is used in the field of gene-altering technology. Because of the force involved, the cells are actually broken, thereby neutralizing the electrical potentials—the very life of the cells—between the outer and inner sides of the cell membranes. Impaired cells become easy prey for microorganisms. Natural repair mechanisms are suppressed, and cells are forced to adapt to a state of energy emergency—they switch from *aerobic* to *anaerobic* respiration.

Instead of water and carbon dioxide, hydrogen peroxide and carbon monoxide are produced. Any reversal of healthy cell processes may occur because of a number of reasons, and our cells then revert from a robust *oxidation* to an unhealthy *fermentation*. The same violent friction and athermic deformations, that can occur in our bodies when we are subjected to radar or microwaves, happens to the molecules in the food cooked in a microwave oven.

A microwave oven decays and changes the molecular structure of the food by the process of radiation. We've all been told that microwaving food is not the same as irradiating it. The two processes are supposed to use completely different waves of energy and at different intensities. In fact, when food is microwaved, the oven exerts a power input of about 1,000 watts or more. This radiation results in destruction and deformation of molecules of food, and in the formation of new compounds (called *radiolytic compounds*) unknown to man and nature.

Microwave cooking changes the nutrients so that significant unhealthy changes take place in the blood of the participants who consume microwaved food—changes that cause deterioration in the human systems.

Cell Phones

The number of people using cellular telephones has risen dramatically during the past decade, and is expected to continue increasing. According to the Cellular Telecommunications Industry Association (CTIA), there are currently (2002) over 110 million wireless telephone users in the United States.

This number is increasing at a rate of about 46,000 new subscribers per day. Experts estimate that by 2005 there will be over 1.26 billion wireless telephone users worldwide. A study by scientists in Finland has found that mobile phone radiation can cause changes in human cells that might affect the brain.

The study at Finland's Radiation and Nuclear Safety Authority found that exposure to radiation from mobile phones can cause increased activity in hundreds of proteins in human cells grown in a laboratory. Nonetheless the study, the initial findings of which were published June 2002, in the scientific journal *Differentiation,* raises new questions about whether mobile phone radiation can weaken the brain's protective shield against harmful substances.

The study focused on changes in cells that line blood vessels and on whether such changes could weaken the functioning of the blood-brain barrier, which prevents potentially harmful substances from entering the brain from the bloodstream.

They also found that one hour of exposure to mobile phone radiation caused cultured human cells to shrink. The researchers believe this is triggered by a response that normally only happens when a cell is damaged. In a person, such changes could disable safety mechanisms that prevent harmful substances from entering the brain from the bloodstream. Radiation-induced changes in the cells could also interfere with the normal death process of *apoptosis*. If cells that are "marked" to die do not, tumors can form. The study found that a protein called *hsp27* linked to the functioning of the blood-brain barrier showed increased activity due to irradiation and pointed to a possibility that such activity could make the shield more permeable, he said. Increased protein activity might cause cells to shrink not the blood vessels but the cells themselves and then tiny gaps could appear between those cells through which some molecules could pass.

The number of people using cellular telephones has risen dramatically during the past decade, and is expected to continue increasing. According to the Cellular Telecommunications Industry Association (CTIA), there are currently over 110 million wireless telephone users in the United States. This number is increasing at a rate of about 46,000 new subscribers per day. Experts estimate that by 2005 there will be over 1.26 billion wireless telephone users worldwide. The concern about an increased risk of cancer with cellular telephone use is related to the radiation that the device produces.

Like televisions, alarm systems, computers, and all other electrical devices, cellular telephones emit electromagnetic radiation. In the United States, cellular telephones operate in a frequency ranging from about 800 to 2400 megahertz (MHz). In that range, the radiation produced is in the form of non-ionizing radiofrequency (RF) radiation. AM/FM radios, VHF/UHF televisions, and cordless telephones operate at lower radio frequencies than cellular phones; microwave ovens, radar, and satellite-stations operate at slightly higher radio frequencies.

RF radiation is different from ionizing radiation, which can present a health risk at certain doses. Ionizing radiation is produced by devices such as x-ray machines and television and computer monitors that use *cathode ray tubes*. Because so many people use cellular telephones, it is important to learn whether RF radiation affects human health, and to provide reassurance if it does not.

A cellular telephone user's level of exposure to RF radiation depends on several factors. These factors include the: amount of cellular telephone traffic, quality of the transmission, how far the antenna is extended, and size of the handset. A cellular telephone's main source of RF energy is its antenna. Therefore, the closer the antenna is to the head, the greater a person's expected exposure to RF radiation. The amount of RF radiation absorbed decreases rapidly with increasing distance between the antenna and the user. The antenna of hand-held cellular telephones is in the handset, which is typically held against the side of the head while the phone is in use.

The antenna of a car cellular telephone is mounted on the outside of the car, some distance from the user. Transportable cellular telephones or "bag phones" have an antenna in a portable unit separate from the handset. Most of the studies conducted on cellular telephone use and cancer risk have focused on hand-held models, since they deliver the most RF radiation to the user. RF radiation can be harmful at high levels because it produces heat. The emitted microwaves heat the metals in the head, such as amalgam fillings braces, crowns bridges, etc.

This increases the galvanic electricity generated as well as the emission of mercury vapour that's inhaled from amalgam fillings. Some people have speculated that the heat produced by RF radiation from hand-held cellular telephones may be associated with brain tumors, because the antenna is held close to the user's head.

Cell phones are becoming increasingly popular, yet most of us are unaware of the damage they are doing to our brains by exposing us to this radiation. Today, there is more and more compelling evidence showing low intensity, pulsed radiation can exert subtle influences that have serious health problems. Your body is an electrochemical instrument of exquisite sensitivity. Its orderly functioning and control are determined by oscillatory electrical processes of various kinds, each characterized by a specific frequency, some of which happen to be close to those used in GSM. Thus, some of your biological electrical activities can be interfered with via oscillatory aspects of the incoming radiation, in much the same way as can the reception on a radio. Your biological electrical activities that are vulnerable to interference from GSM radiation include highly organized electrical activities at a cellular level, whose frequency happens to lie in the microwave region, can interfere with processes as fundamental as cell division.

This radiation also seems to affect a variety of brain functions including the neuroendocrine system. Additionally, cell phones have been associated with a 200-300% increase neuroepithelial tumours that occur on the same side of the brain that the cell phone was being used on. People should limit their exposures to these potential dangerous devices that cause irreversible brain damage and cancer.

We only have to look less than a hundred years back to learn from similar mistakes our forefathers made in this area. Early in the 20th century radon and radium-enriched spa waters were "recommended" for a wide range of aches and minor ailments. Today, no reputable scientist would disagree with the dangers that ionizing radiation has to our health.

Unfortunately it is nearly technically impossible to design a study to prove that a device or procedure is safe. That is one of the reasons that cell phones health hazards have not received more attention than they have. Cell phones have been one of the fastest growing industries in modern history. 300 million Americans now use cell phones, and this technology is predicted to have over 1.3 billion worldwide users by the year 2005.

Every day some 300 million cell phone users are "reaching out and touching someone" they love with the new 1.9 gigahertz (nineteen-hundred-million) broadcasting frequency. Yourself, and anyone else within range of the microwaves emitted by your phone. The uses of electronic communications for average people began with the introduction of personal paging systems in the 1970's—expanding into remote telephones and cell phones by the end of the century. Most people today have either portable phones in the home, cell phones for away-from-home use or both. As entire nations reach for pocket communicators, the explosively emergent $40 billion a year mobile phone industry is poised to deliver a "Wireless Revolution" that over the next five years is expected to double the one-billion people connected by telephone lines over the past century.

THE HEALTH EFFECTS MOUNT

Exquisitely sensitive to subtle electromagnetic harmonies, human brains and bodies as intricate as galaxies depend on tiny electrical impulses to conduct complex life-processes-including the ability to read, recall and respond to these words. Much like a boxer taking repeated blows to the head, rapidly pulsing cell phones signal permanent brain damage. Acting as antennas, our anatomies just as easily tune into spurious signals from radio and microwave transmissions. When it comes to cell phones, a worse frequency could not have been chosen for the human anatomy. The brain reaches peak absorption in the UHF bands, right where cellular telecommunications operate.

A study by Dr. Peter Franch found unequivocally that "cells are permanently damaged by cellular phone frequencies." This cellular damage, Franch noted, is maximized at low dosage, and "inherited unchanged, from generation to generation." Franch found that the production of *histamine*, which triggers bronchial spasms, is nearly doubled after exposure to mobile phone transmissions.

Cell phones also reduce the effectiveness of anti-asthmatic drugs, and retard recovery from illness. British military scientists have discovered that cell phone transmissions disrupt the brain sites for memory and learning, causing forgetfulness and sudden confusion. Other studies show that electromagnetic signals from cellular phones reduce the ability to concentrate, calculate and coordinate complicated activities such as driving a car.

Mobile telephones are arguably the most radiative

appliance we have ever invented apart from the microwave oven and people are putting them by their heads—arguably the most sensitive part of the body. The pegged needle of a shrieking electromagnetic radiation (EMR) meter placed beside a connected cell phone still shows significant exposure 100 feet away.

Cell phones emanate microwave radiation, and human brains may absorb up to 60 per cent of that energy. Already there are at least 15,000 scientific reports on the subject. It has been discovered that small amounts of energy, when delivered in the right way, can have the same effects as a massive dose of chemicals. The most delicate fields of electromagnetic energy influence the complexity of living creatures. There are 66 epidemiological studies showing that electromagnetic radiation across the spectrum increase brain tumors in human populations. Two of those studies are for particular brain tumors from cell phones. Because cancer takes decades to develop, it will be another 10 or 20 years before "mobiles" manifest a bonanza in brain tumors. We're already seeing acute effects that are noticed within minutes of using a cell phone.

After two minutes' conversation, a cell phone's digitized impulses disable the safety barrier (*blood-brain-barrier*) that isolates the brain from destructive proteins and poisons in the blood. Professor Leif Salford, the neurologist who carried out the research for this finding, said "It seems that molecules such as proteins and toxins can pass out of the blood, while the phone is switched on, and enter the brain. We need to bear in mind diseases such as MS and Alzheimer's are linked to proteins being found in the brain." Startled by $4 billion a year in extra claims among cell phone-wielding drivers, North American insurers did a double-take that found simply juggling phones is not causing a 600% increase in accidents over other drivers busy shaving, applying makeup, tuning radios, taming pets, making out, pouring coffee, retrieving dropped cigarettes, talking and gesturing to passengers, or actually steering the vehicle.

Instead of just another dangerous distraction, tests conducted by the U.S. Department of Energy found that using a cell phone severely impairs memory and reaction times. Hands-free mobile speakerphones cause even more crashes because they typically emit 10-times more brainwave interference than handheld units. For all drivers dialing out, University of Toronto investigators report that the heightened probability of cracking up your car persists for up to 15-minutes after completing a call.

That's comparable to the risk of crashing while driving dead drunk exclaims Dr. Chris Runball, chairman of the B.C. Medical Association's emergency medical services committee. Reeling from "dial-a-collision" costs, the government of British Columbia may join England, Spain, Israel, Switzerland and Brazil in restricting or banning the use of cell phones by drivers. In New Zealand, cell phone towers are prohibited on school property because of possible health effects.

But Health Canada regulations ignore the hidden hazards of cell-wrenching cell phones, which send pulsed signals through the skull in a process one expert likens to "jackhammers on the brain." "Safety Code 6" looks only at microwaves burning skin.

Basically, Health Canada claims if it can't cook you, it can't hurt you, its like saying cigarettes aren't dangerous unless they burn you. Carrying similar risks of long-term lethality, and strangely just as legal, cell phone addiction mirrors the prestigious early allure of smoking–as well as an immensely profitable industry's steadfast denial of risk and responsibility.

As poisonous as cigarette smoke and even harder to corral, the cell phone's second-hand microwave and radio-frequency (RF) pollution pose invisible but significant risks to bystanders—particularly children riding in cars that transmit amplified cell phone signals through their steel structure. Cell phone transmissions damage the ability of white blood cells to ward off infectious disease by disrupting the immune system's electromagnetic communications.

In 1995, mobile phone sales in North America exceeded the birth rate. Hired by the Cellular Telecommunications Industry Association to condone cell phones, public health scientist George Carlo found that rare tumors on the outside of the brain are more than doubled among cell phone callers—particularly on the right side of the head where phones are usually held. Carlo told ABC's "20/20" that cell phone causes genetic damage that leads to cancer. Warning of "the potential for a global health disaster," ABC recommended "prudent avoidance" of cell phones after finding that every cell phone they lab-tested exceeded the Federal Communication Commission's standards for EMF absorption rates.

EMF researcher Dave Ashton cautioned 20/20 viewers that because cell phones constantly search for the nearest repeating tower, "long-term damage comes from cell phones in the stand-by mode." Cell phone "shields" and headsets cannot adequately address these problems. Dr. Carlo later told London's Express newspaper that cell phones cause genetic damage following a dose-response curve. That is, the more a person uses a cell phone, the more cellular destruction and health risks they incur.

Cell phone-confused cells can go crazy, Carlo cautioned. Experiments on captive animals show that this cumulative DNA damage is passed on to succeeding generations. Addicted as we are to a culture of convenience, we forget how inconvenient it is to contract cancer. As magazine-size "cellular" relay antennas hidden in church steeples and rooflines keep popping up just about everywhere, more and more communities are declaring their airspace a "No Fry Zone."

Cellular relay towers in Kansas and Oklahoma have been shut down because they interfered with passing aircraft. These same transmissions pluck our electrical circuitry, inflicting a "chronic edginess" that "twangs human nerves." Sleep disorders are common among people exposed to high levels of electromagnetic pollution. These energy fields can affect the central nervous system and immune system in man. Chronic exposure to EMFs is associated with significant changes in some physiological parameters.

The impact on Calcium ions, which are important in maintaining normal health functions in brain tissues, was found in experiments. This may affect nervous system function. Amplitude-modulated (AM) radio frequency radiation can induce responses in cells of nervous tissue origin from widely different animal species, including humans. Another area of concern is in prenatal development in mammals. Australian research has found one of the strongest links between cell phones and cancer. Over periods of 9 to 18 months, exposed mice had twice the tumor rate as unexposed mice. The mice were exposed to cell phone radiation.

As reports linking cell phone use to adverse health conditions have been published, attempts have been made by industry to hose down the findings with what is called "The Hockett Defence" (named after the chief Tobacco Institute scientist) who advise his executives to repeat endlessly, "men aren't rodents." But, DNA is DNA. At the level of normal cell growth processes, human and animal cells act very similar. The body is not a closed system but, rather, an open one, which exchanges energy with all of the forces around it. The human body seeks equilibrium or its own balance.

Energy interaction requires a corresponding action from the body. On an energetic level this results in chemical changes, system stress and other interactions, which can be either healthy or not so healthy. EEG-data of man sampled under the influence of these electromagnetic fields are altered extremely in the range of alpha-activity *during* as well as *after* exposure for some hours.

This biological effect is induced by field intensities lower than the given international limiting values. Regarding these results there is the very important question of possible influences, injurious to health for people exposed to pulsating electromagnetic fields, especially by operating the new type of digital telecommunication networks.

Even fairly low levels of electromagnetic radiation can change the human body's sleep rhythms, affect the body's cancer-fighting capacity by harming the immune system, and change the nature of the electrical and chemical signals communicating between cells. Reports continue to amplify the same results, which are being replicated now around the world.

At the same time the industry is shifting the standards, changing cell phone designs and altering other factors that make evaluation of the effects even more difficult. Existing data indicate that RFR of relatively low intensity can affect the nervous system. Changes in blood-brain-barrier, morphology, electrophysiology, neurotransmitter functions, cellular metabolism, and calcium efflux, and genetic effects have been reported in the brain of animals after exposure to RFR. These changes can lead to functional changes in the nervous system. Behavioral changes in animals after exposure to RFR have been reported.

New research indicates that exposure to cell phones' radiation causes red blood cells to leak hemoglobin, the build up of which can cause heart disease and kidney stones. Even at lower levels than those emitted by cell phones, the blood cells leaked hemoglobin.

Two minutes of exposure to emissions from mobile phones can disable a safety barrier (blood/brain barrier) causing proteins and toxins to leak into the brain. This can cause the chances of developing diseases such as Alzheimer's, multiple sclerosis and Parkinson's. Symptoms reported by mobile phone users include fatigue, dizzy spells and memory loss. Amid an explosive growth of mobile communications, concerns are mounting about cellular telephones' potential links to health problems ranging from headaches to brain tumors. British scientists are demanding that mobile telephones carry a health warning. The National Radiological Protection Board (NRPB) recommends a limit of 10 milliwatts per gram, though proposed European guidelines are five times more restrictive. There have been reports of headaches caused by cell and portable phone use. Evidence from the 1960s and 70s supports the conclusion that cell phones cause headaches among some users.

Cell phone transmitting frequencies fall in the most sensitive band for the microwave hearing effect. The transmitting frequencies are also in the band that has maximal penetration into the head. The area of the head that needs to be exposed to the microwaves in order for people to perceive the effect is in proximity to the antenna of present day cellular telephones. 30 years ago Dr. Frey encountered and reported headaches from microwave energy exposure at approximately the same frequencies, modulations and incident energies that present day cellular telephones emit.

Low intensity microwave exposure opens up the blood/ brain barrier, a biological effect that can allow the release of dangerous chemicals into the brain. The radiation from mobile phones makes it easier for poisons to penetrate into the brain. Data in the literature now indicates that the dopamine/opiate system may be involved in headaches and suggest that headaches may be due to cellular emissions...the energy used was approximately the same in frequencies, modulations, and incident energies as those emitted by present day cellular telephones.

These current reports of headaches may be the canary-in-the-coal mine, warning of biologically significant effects. The early research by Dr. Frey was most revealing in that it was conducted before the advent of the cell phone. As a result, the research was done with limited, if any, economic impact on industry and the results were unchallenged. Exposure to electromagnetic fields during mobile phone use may increase resting blood pressure.

Exposure of the right hemisphere to a radio-frequency electromagnetic field for 35 minutes causes an increase in sympathetic efferent activity with increases in resting blood pressure between 5 and 10 mm Hg, most likely due to more pronounced vasoconstriction.

Mobile phones can cause sudden confusion and short-term memory loss, according to worrying research by British military scientists. Signals from the phones disrupt part of the brain, which controls memory and learning. Frequent mobile users had significantly depressed melatonin–a vital cancer-preventing hormone. An Australian study has linked the phones to a higher rate of brain cancer while a Swedish survey suggested that using a mobile phone for more than 15 minutes could lead to headaches and fatigue.

Exposure to RFR does produce various effects on the

central nervous system. Other parameters of RFR exposure, such as frequency, duration, waveform, frequency- and amplitude modulation, etc, are important determinants of biological responses and affect the shape of the dose (intensity)-response relationship. As we have increased our exposure to both chemicals and EMFs in the last three decades we have seen certain brain cancers increase in all age groups by 1% a year. Since 1980 those 65 and older have increased 2.5% a year. Among those over 85 the increase has been as much as 500% since 1973.

If you must pack a cell phone, treat it like a loaded pistol. Keep it turned off. Don't carry it near ovaries, testicles, or the heart. For partial protection, buy an antenna shield. Limit calls to one-minute, six to 10 minutes a month. Never fire off a cell phone with children anywhere in sight. A better bet is to facilitate the growth of organic telephone networks with lots of fibre. Instead of more microwave towers, we should be wiring up our cities with fibre-optic cables to provide Internet, fax, telephone, radio and television at very high quality, rather than saturating our cities with the microwave, radio wave and low frequency signals all the time.

When it comes to cells, consciousness and cell phones, every call is collect. How can convenience count more than cancer? What is gained by being in constant contact with disembodied voices, while being "out of touch" with the friends and neighbors around us? Are we comfortable having our location traced by monitoring authorities? Unless we start voting with our wallets, consumer complacency could prove as species-limiting as corporate cynicism.

Microwave frequencies are the same as those used in radar and your microwave oven. You wouldn't think of sticking your head in the microwave oven, but there is no hesitation to putting the cell phone to your ear.

Remote Home Portable Phones

Cell phones are not the only phones to cause concern about health problems. Today's high-frequency cordless phones may emit a level of EM radiation similar to cell phones.

For reasons of health and privacy, a growing number of scientists and other experts are dead set against these phones. They say a cancer risk is associated with signals that have a strong wattage and high frequency (short wavelength). First generation cordless phones operate at about 60 megahertz and the next ones ran at 900 megahertz—higher than 835MHz cell phones. The new 2.4 GHz is higher still and can transmit for several kilometers without fading. Microwave ovens operate at 2.45 GHz.

Brain Tumors

Two new studies have shown links between mobile phone usage and brain tumors. They found a correlation between phone use and cancer. The studies showed that mobile phone use, regardless which side of the head it was held against, increased the risk of a brain tumor by almost two and a half times. Some of the leading researchers were contracted by Motorola to carry out some experiments.

Two of the world's leading radiation experts reported to *The Express* that multinational companies tried to influence the results of their research. Professor Ross Adey, a biologist, had his funding withdrawn by Motorola before completing research which showed that mobiles affected the number of brain tumors in animals.

Dr. Henry Lai, who has been studying the biological effects of electromagnetic fields for 20 years, was asked three times to change findings on how they caused DNA breaks in rats. Both of these scientists have been involved in academic, military and other research projects throughout their lives.

Jerry Phillips, who has a doctorate in biochemistry, worked with U.C. Riverside's Adey on Motorola-funded research beginning in 1991. He describes a pleasant relationship with them until studies linked exposure to changes in the incidence of brain tumors in rats. "Motorola was adamant that Adey never mention DNA damage and RF radiation in the same breath," Phillips said. "Motorola has been manipulative of research that we and others have reported to them," said Adey. "Essentially they cut us off because we were too inquisitive." Phillips, Adey, and others said they see a strong parallel between what's happening now and the decades of denial by the tobacco industry.

Though the head absorbs 60% of the energy radiated from cell phones, the amount is not significant enough to cause heating. Evidence points to DNA damage as a source of the health problems associated with cell phone use. It is suggested that RFR may hinder the ability of DNA to repair itself. For the first time in history, we are holding a high-powered microwave transmitter against the head.

When you talk on your mobile phone, your voice is transmitted from the antenna as radio frequency radiation (RFR) between 800MHz and 1,990MHz at a range that's right in the middle of microwave territory. Public exposure to electromagnetic radiation (radio frequency and microwave) is growing exponentially worldwide with the introduction and use of cordless phones, cellular phones, pagers and antennas in communities designed to transmit their RF signals.

The virtual revolution in science taking place now is based on a growing recognition that non-thermal or low intensity RF exposure can be detected in living tissues and result in well defined bioeffects, including changes in cell membrane function, metabolism, cellular signal communication, activation of proto-oncogenes, and cell death.

Resulting effects which are reported in the scientific literature include DNA breaks and chromosome aberrations, increased free radical production, cell stress and premature aging, changes in cell membrane function including memory loss, learning impairment, headaches and fatigue, sleep disorders, neurodegenerative conditions, reductions in melatonin secretion, and cancer. The United States has a de-facto policy of "post-sales surveillance" with respect to cell phones. That means cell phones can be sold to the public, and only after years of use will there be studies to characterize what health consequences, if any, have arisen as a result.

Some scientists are even warning that constant cell phone use causes premature aging. Low level radiation from the phone "heats up" body cells, damaging skin and making the user look lined and haggard. Gradually, cells don't work properly, so the life process becomes less efficient. Heavy mobile phone users are just like heavy smokers who constantly inhale cell-damaging toxins without allowing the body time to repair the harm.

The Effect on Children

The fastest growing group of mobile phone users are children and young people. This growth is actively encouraged by professional advertising campaigns from the mobile phone industry, extolling how indispensable the phones are to their lifestyles. They have really gone out after the young people with prepay cards and colored handsets. Phone makers often present their products in adverts as essential "back to school" items for children.

This advertising blitz was produced by the same transnational public relations corporations that previously gave us such delightful cartoon characters as "Joe Camel" for the tobacco industry, and no words of warning are heard. However, within the scientific community, there is a growing chorus of expert voices that are urging caution because, if there are adverse health effects from mobile phone use, it will be the children who will be in the front line and who may pay the highest price.

For the sake of the future of our children's health, we need to heed these voices seriously and limit children's unnecessary use of mobile phones. Children may be more vulnerable because of their developing nervous systems, the greater absorption of energy in the tissues of the head, and a longer lifetime of exposure. Children are more sensitive to microwaves than adults; they absorb microwaves at 3.3 times the rate of adults. While some manufacturers target children for cell phone sales, cell phone radiation penetrates the skulls and brains of kids more deeply than adults, and this radiation might cause tumors or otherwise affect a developing brain.

Children are more vulnerable because of their developing nervous system, the greater absorption of energy in the tissues of the head, and a longer lifetime of exposure. Absorption of microwaves of the frequency used in mobile telephony is greatest in an object about the size of a child's head, the so-called *head resonance*, while in consequence of the thinner skull of a child, the penetration of the radiation into the brain is greater than in an adult.

Children's brain wave patterns are abnormal and stay like that for a long period. This could affect their mood and ability to learn in the classroom if they have been using a phone during break time, for instance, leading to things like a lack of concentration, memory loss, inability to learn and aggressive behaviour. The still-developing nervous system and associated brainwave activity in a child are more vulnerable to aggression by the pulses of microwaves used in these devices than is the case with a mature adult.

This is because the multi-frame repetition frequency of 8.34 Hz and the 2 Hz pulsing that characterizes the signal from a phone equipped with discontinuous transmission (DTX) lie in the range of the *alpha* and *delta* brainwave activities, respectively. The fact that these two particular electrical activities are constantly changing in a child until the age of about 12 years, when the delta waves disappear and the alpha rhythm is finally stabilized, means that they must both be anticipated to be particularly vulnerable to interference from the pulsations. The increased *mitotic* activity (cell division) in the cells of developing children makes them more susceptible to genetic damage.

A child's immune system, whose efficiency is, in any case, degraded by radiation of the kind used in mobile telephony, is generally less integrous than that of an adult, so the child is less able to cope with any adverse health effect provoked by chronic exposure to such radiation. Scientists have discovered that a call lasting just two minutes can alter the natural electrical activity of a child's brain for up to an hour afterwards. They also found for the first time how radio waves from mobile phones penetrate deep into the brain and not just around the ear.

Doctors fear that disturbed brain activity in children could lead to psychiatric and behavioural problems or imair learning ability. Several brands of cell phones exceed the radiation limits specified in FCC guidelines, ABC News show *20/20* reported. The widespread use of mobile phones by children for non-essential calls should be discouraged. The mobile phone industry should refrain from promoting the use of mobile phones by children. Epidemiology has seen a large number of examples where health hazards were initially described with unconvincing and sometimes inadequate experiments that demonstrated a weak association with a given environmental influence.

Such associations were found between cholera and drinking water containing fecal contaminants, between smoking and lung cancer or between exposure to vinyl chloride and certain forms of liver cancer. All these associations were highly questioned in the past and are now well recognized.

Current Research on Cell Phones

Though 85 million Americans now use cell phones, Europeans began widespread use of them much earlier with many now reporting side effects from their use. At least one of the symptoms noted, which include dizziness, concentration difficulties, memory loss, and a burning sensation, showed up in 47 per cent of people who reported using these wireless devices an hour or more daily.

The China Consumers Association (CCA) issued a "worrisome" warning about Chinese cell phones' electromagnetic radiation and how this might affect phone users. According to the CCA, tests have found that some cell phones' radiation was as high as 10,000 microwatts per square centimeter (1,550 mw per square inch). Half of cell phone's radiation is absorbed by the human body and another quarter by the brain.

The strongest link yet shows a high cancer death rate among soldiers exposed to microwave radiation, the same as that emitted by the phones. Using a mobile phone could drastically reduce your sex drive, new research shows. Exposed rats had far less testosterone in their blood stream than those that remained unexposed. The higher the dose of radiation, the less testosterone was released by the body's glands, resulting in diminished sexual activity." The test results are being studied in Britain, where scientists are conducting similar research.

Current safety guidelines for cell phones assume no harmful effects, as long as the microwave radiation they emit does not cause heating of body tissue. Exposure limits are intended to protect us only from excessive temperatures caused by absorption of energy, a known danger linked to the intensity of radio frequency microwaves. But living cells respond in non-thermal ways to the fields produced by cell phones, and at intensities below the established safety threshold.

Ionizing radiation is a well-established risk factor for brain tumors. Exposure to ionizing radiation, work in laboratories, and work in the chemical industry increased the risk of brain tumors. Use of a cellular telephone is associated with an increased risk in the anatomic area with highest exposure. In addition to extremely low frequency (ELF) electric power fields, many millions of mobile phone users worldwide are now also exposed daily to radio frequency fields under near-field conditions.

We may expect that these newly evolved behavioral patterns will be lifelong, with intermittent exposures at the phone user's head making yet one more contribution to an already complex daily EMF exposure arising in an aggregate of multiple and disparate sources. The cell phone industry continues to downplay the risk and defer to more research.

The precautionary principle is a good one for an age in which technology is doubling every few months and the impact of that technology may not be known for years after. We need to open the doors in this area and in the mean time, find ways to reduce risk and exposures for EMFs—particularly those created by home remote and cell phones.

Independent Research Funded by Industry

The industry has been involved in attempting to influence

the research and has been required to pay for independent research. The companies involved have attempted to control the data flow to the public as information has become available.

As far back as 1996 the issues began to surface with the following report. Motorola Inc. planned how to collaborate with the Cellular Telecommunications Industry Association and Wireless Technology Research L.L.C. to downplay potentially damaging scientific findings on possible health risks from portable telephones, according to a Dec. 1994 internal Motorola memo.

The company the industry sought to collaborate with was the firm, which eventually used over $27,000,000 industry dollars to research the risks of cell phones. In the body of that research several issues emerged creating additional studies.

The U.S. wireless industry, responding to the global proliferation of media coverage of mobile phone health concerns and to Internet-savvy activists, is leading an effort to create a worldwide information-sharing network to counter negative publicity. The Wireless Industry Global Information Network, or WIN, held its first meeting Dec. 10, 1998 in London. Dr. George Carlo's initial reports were framed in the standard conservative approach at finding the facts which science demonstrates.

In an attack on the industry for which he once acted as a spokesman, he accused firms of not taking safety seriously. Dr. George Carlo, in his capacity as director of Wireless Technology Research wrote a letter to the CEO of AT&T, which has serious legal implications for mobile phone manufacturers who have claimed that there is no evidence for adverse health effects from mobile phone use. With the letter widely circulated in the industry, making that claim now could possibly expose them to litigation in much the same way as what happened to the tobacco industry, where it was shown that industry assurances of no evidence of hazards from smoking was a complete fabrication.

The rate of death from brain cancer among handheld phone users is higher than the rate of brain cancer death among those who used non-handheld phones that are away from their head. The risk of *acoustic neuroma,* a benign tumor of the auditory nerve that is well in the range of the radiation coming from a phone's antenna, was fifty per cent higher in people who reported using cell phones for six years or more, moreover, that relationship between the amount of cell phone use and this tumor appeared to follow a dose-response curve.

The risk of rare *neuro epithelial* tumors on the outside of the brain was more than doubled, a statistically significant risk increase, in cell phone users as compared to people who did not use cell phones. There appeared to be some correlation between brain tumors occurring on the right side of the head and the use of the phone on the right side of the head. Laboratory studies looking at the ability of radiation from a phone's antenna to cause functional genetic damage were definitely positive, and were following a dose-response relationship.

The cell phone transmits a microwave signal from an antenna to a base station or tower, often miles away. The farther from the tower, or if the phone is inside a building or a car, the more power this phone is told by the tower to send out to make or keep the connection.

Depending on how close the cell phone antenna is, as much as 60 per cent of the microwave radiation is absorbed by and actually penetrates the area around the head, some reaching an inch to an and a half into the brain. The $200-billion-a-year cell phone industry maintains the devices are safe. The question of cell phone safety recently led Metrocall of Alexandria, Va., the nation's third-largest pager company and a major seller of AT&T cellular phones to warn its sales staff that parents buying for a child or young adult should consider a pager instead of a cell phone due to potential health risks.

Liability and Possible Claims

The industry has largely put forth studies that looked at the effects of radio waves outside the cellular frequency, or at exposure levels that are different from those experienced by cellular phone users. The industry hasn't told the public the full story about how there has been very little research on biological effects at low-level exposures, similar to those of handheld phones. Very limited information has been available to the public about the risks of cell phones or various electromagnetic fields outside of some obscure research and academic circles.

The fact is that increasing evidence has been mounting and the true risks of these energy fields are becoming well known. The possibility of another tobacco-type of health scandal was perhaps in the offing when a closed Congressional hearing was held to develop regulations and recommend further studies of electromagnetic field (EMF) health effects.

They suggested moderation in phone use until more is known, while an FDA suggested: those who spend long periods of time on their hand-held cellular phones could consider holding lengthy conversations on conventional phones and reserving the hand-held cellular models for shorter conversations. Many studies have been sponsored by industry, academic institutions, and government laboratories and by military research organizations into the effects of low levels of electromagnetic radiation.

The constant problem in the debate of risks is the limited knowledge about very specific fields, interacting with our bodies, having significant health effects. These effects vary throughout populations with some effected to a greater degree than others. This is related to our physical and biochemical differences. The research that is being conducted by the industry is ignoring much of what has already appeared in the literature regarding risk factors.

The FDA concluded in a February 8, 2000 report that; There is currently insufficient scientific basis for concluding that wireless communication technologies are safe or that they pose a risk to millions of users. A significant research effort, including well-planned animal experiments, is needed to provide the basis to assess the risk to human health of wireless communications devices. The FDA began a three to five year study to look at some of these effects. This came at the conclusion of the industry sponsored study which, along with other recent studies, show increasing risk to human health related to cell phone emissions. The act amends the Communications Act by providing federal preemption of state and local regulation of personal wireless service facilities on the basis of RF environmental effects. In other words states and local communities may not adopt more stringent protections if the federal regulatory authorities fail to protect the public.

This limitation on the rights of states or local governments essentially leaves any risk assessment and solution to the federal level regulatory authorities. The FDA approach and the reluctance of the United States government to move on this issue is directly related to lobby efforts, public relations gimmicks and the manipulation of the facts behind what is a major concern to many. Other governments are taking a different approach.

The risks associated with cell phones are being considered too risky even by the biggest risk takers in the insurance industry. Concern about the safety of mobile phones has prompted a leading Lloyd's underwriter to refuse to insure phone manufacturers against the risk of damage to users' health. Fears that mobile phones will be linked to illnesses such as cancer and Alzheimer's have prompted John Fenn, of underwriting group Stirling, to refuse to cover manufacturers against the risk of being sued if mobiles turn out to cause long-term damage.

Risk management and the kinds of legal concerns arising out of the tobacco lawsuit has turned insurers of product liability claims on their heads. A number of employers expect their employees to carry out their duties and responsibilities using mobile phones for hours at a time. It could well turn out to be a non-safe system of work for which substantial damages may be awarded as a result of adverse health conditions. A number of cases have already been settled out of Court but again subject to confidentiality clauses. Companies recognizing potential third party litigation have also been attempting to reduce their risk.

Employers who require cell phone use in the course of employee work are putting forward reasonable technological advances, which could decrease risk. The risk of future litigation increases if the precaution is not taken to provide a safe workplace. The removal of the cell phone from the head is helpful to reducing exposure to the head but it can still expose the body to the effects of its radiation.

Base-stations or cell phone towers also pose risks. The installation of base stations for mobile telephone systems has been delayed or has met opposition from the public because of concerns that the RF emissions from these base stations might cause cancer in children. In the United States, for example, 85% of the total number of base stations needed have yet to be constructed. The significant increase in these systems and their interactions with other energy fields in our homes, cars and work places will in fact be significantly increasing health risks.

International organizations are also looking at the risks because of increasing public concern, scientific evidence and industry concerns. In May 1996, in response to growing public health concerns in many member states over possible health effects from exposure to an ever-increasing number and diversity of EMF sources, the World Health Organization (WHO) launched an international project to assess health and environmental effects of exposure to electric and magnetic fields, which became known as the International EMF Project. Other studies are already producing the evidence of biological effects.

The research continues and the health effects mount. With over 1.3 billion people projected to be using these devices in the year 2005 the risks must be understood and addressed.

The FDA advises persons concerned about exposure to cell phone radiation to take some simple steps to avoid exposure. Those persons who spend long periods of time on their hand-held mobile phones could consider holding lengthy conversations on conventional phones and reserving the hand-held models for shorter conversations. People who must conduct extended conversations in their cars every day could switch to a type of mobile phone that places more distance between their bodies and the source of RF, since the exposure level drops off dramatically with distance.

How cell phones might effect our health is compelling people to take heed of the warnings and find ways to reduce exposure without giving up the device. Some have resorted to earpieces and belt level phones. This may in fact prove to be even worse than head exposure because of the way in which the phone then operates. More power is required and exposure to softer tissue allows more energy to transfer into the body.

Cars and other vehicles screen the microwave emissions from the mobile phone when in use and so act as a *Faraday cage*. GSM phones compensate for this by increasing the power output resulting in greater microwave absorption in the user.

CELL PHONES MAKING HONEYBEES DISAPPEAR

The findings could provide an answer to the mystery of disappearing bee colonies across the Western world. Scientists are having a difficult time explaining the rapid decline of the honeybee population in the United States and Canada, known as colony collapse disorder, that could affect crops depending on these insects for pollination. Research presented some four years ago speculates the radiation emitted from cellular phones may contribute to the problem. When cellular phones were placed near hives, the radiation generated by them (900-1,800 MHz) was enough to prevent bees from returning to them, according to a study conducted at Landau University.

Scientists believe the radiation produced by cellular phones may be enough to interfere with the way bees "communicate" with their hives. Cellular phones may create a resonance effect that interferes with the movement patterns bees use as a kind of language.

The cell phone study, conducted by Landau's Jochen Kuhn, was originally presented at a German Informatics and Cybernetics conference back in 2003. It focused on the effects of cell phone radiation on the neurological mechanisms that control learning and memory. Placing handsets near hives, Kuhn observed that GSM cell phone radiation in the frequency range 900 MHz-1800 MHz caused the bees to avoid the hive. Kuhn speculates that the "waggle" dance that bees perform on the honeycomb to communicate with others could be influenced by the radiation.

Although another study blamed the ongoing spread of genetically modified crops for the disappearance, cellular phones may also very well be contributing to the problem. This comes as no surprise, considering all the harm mobile phones can do to male sperm, for example, as well as to your head.

Not all scientists agree that the dance is the primary means of communication between bees, but past studies have shown that it can fundamentally affect bee movement patterns. Kuhn suggests that the 200-300 cycles/second oscillations that dancing bees produce through honeycomb may be interrupted by a resonance effect caused by the telephone handsets.

Hands-free Kits

Hands-free kits are revealing the hidden risks it turns out that "hands-free" cell phones may not save you from the Grim Reaper after all. Currently, the cellular industry is encouraging the use of hands-free kits, but cite their only advantage as being to offer greater freedom of use to the user, nothing else. The fact is, they know the real reason users are buying them is because they think these kits protect them from radiation exposure. But, hands-free kits are no guarantee of lowering radiation emissions from mobile phones, and, in some cases, they actually increase it.

Recent tests have shown that while direct exposure to the head is reduced by around 70 per cent, all that is happening is the radiation is being transferred to another part of the body, which is potentially more vulnerable because it does not have the thickness of the skull to protect itself, such as the waist or chest areas. Earpieces serve as antennae that direct more electromagnetic radiation into the ears.

Phone models may differ in the degree to which their internal wires conduct electricity. Not only are many hands-free devices useless in protecting wireless phone users from radiation that might cause tumors, these products may actually raise the amount of radiation being directed into the head by three times. Using a mobile phone clipped to your waist results in a hotspot of radiation being pumped into the liver and kidneys, prostate, uterus, and ovaries. There is concern that they may intensify radiation exposure to the ear canal. Using a hands-free kit and making a call with a mobile phone clipped to your belt also means the phone will generally be working at a higher power level. That's because it is generally harder to transmit from waist-height than head height. But there's a lot of body tissue in that area which has good conductivity and absorbs radiation more quickly than the head. People think hands-free kits are safer, so tend to spend more time on the phone. Ii works harder to pick up a signal if it is down at waist, where more radiation is absorbed, than at the head.

Non-thermal Verses Thermal Effects

The literature is full of published papers showing damage and biological effects at power levels that were set deliberately well below thermal thresholds and therefore by definition could not have been caused by thermal damage. Even if we assume the thermal only argument to be correct, there are circumstances that the group should be aware of, where exposure exceeds even the thermal limits.

Mobile users who wear metal-rimmed glasses are intensifying the exposure to their eyes by 20 per cent and into the head by 6.3 per cent. Using a mobile in a vehicle can accelerate radiation levels by up to 10-fold due to resonance effect. These risks associated with remote telephone use cannot be ignored. Intelligence documents show that Western governments have known about Soviet experiments using mobile phone-type radiation to cause brain damage for more than 20 years.

The uncensored documents reveal that Soviet military scientists had successfully used microwaves of the type used by mobile phones to weaken the blood brain barrier. US Army scientists had succeeded in duplicating the Soviet experiments by 1977—eight years before mobile phones became generally available in Britain. This work was done as a result of microwave bombardments of the United States Embassy in Moscow as well as reports about research by the Soviets.

At that particular time the safety standards for exposure to radio frequency radiation in the Soviet Union was significantly more stringent than United States standards by almost 1000 times. Russian and other East European countries' exposure standards for radio frequency and microwave radiation are much stricter that in the U.S or Western Europe.

Western standard setting organizations have emphasized protection from RF/MW thermal effects...while Russia's more restrictive standard also reflects a concern over non-thermal effects and subjective symptoms. Biological studies of enzymes and human cells exposed in vitro to radio frequency/ microwave fields have shown a number of effects which cannot be explained simply by the heating effects of radiation on which our current standards are based.

These include changes in cell membrane permeability to potassium, sodium and calcium; changes in the composition or behaviour of blood-forming and immunological cells; alteration of calcium ion exchange in nerve tissue; changes in the firing patterns of neurons; and changes in levels of cancer related enzymes. Very close range exposure to microwaves from a cellular phone base station increases the effect of a chemical mutagen on human blood cells, leading to increased chromosomal aberrations. Microwave radiation such as that emitted by cell phones can heat objects in its path. The heat generated causes the body to begin to expend energy to cool the area and otherwise return to its state of equilibrium before it was irradiated by the device. One study, by the American Health Foundation, in New York, found that mobile phone users had double the risk of developing a certain type of brain tumor than people who did not use them.

A second study found that DNA in human blood cells broke down when exposed to large doses of mobile-phone radiation, possibly laying the genetic groundwork for cancer. Most studies have examined the results of short-term, whole body exposure to RF fields at levels far higher than those normally associated with wireless communications. With the advent of such devices as walkie-talkies and mobile phones, it has become apparent that few studies address the consequences of localized exposures to RF fields to the head.

Scientists have reported other effects of using mobile phones including changes in brain activity, reaction times, and sleep patterns. When mobile phones are used close to some medical devices (including pacemakers, implantable defibrillators, and certain hearing aids) there is a possibility of causing interference. There is also the potential of interference between mobile phones and aircraft electronics.

Reducing the Risk

As the EMF/EMR health effects issue becomes more widely known, especially in relation to mobile phone use, there is a corresponding increasing number of so called EMF protective devices being advertised in health and alternative magazines as "cure-alls" which apparently claim to provide complete protection from exposure to all forms of man made electromagnetic fields (EMF). There is a wide range of devices being offered that make all kinds of unsubstantiated claims.

- Reducing use is universally regarded as the best step. Use by children should be eliminated. Indoor use increases exposure significantly because the signal strength requires creating a connection from inside a car or building is much greater. Home portable phones should be replaced with the old style hard-wired phones and cell phone use significantly reduced.
- If still using a cell phone or portable home phone keep the phone away from the body when in standby mode. When in use hold the phone as far away from the head as possible. Even three or four inches can significantly reduce the exposure because the energy density drops very rapidly with distance from the body. Keep the antenna away from the head and pointing away from the body. Radiation from all sources obeys the inverse square law. That is, the further you are from the source the less intense your

exposure to the radiation. In fact, it drops off with the square of your distance from the source. If you are twice as far from a fire you feel 1/4 of the radiant heat, but if you move four times as far away you only feel one-sixteenth of the heat.

- None of the devices claiming to eliminate all emissions have any science behind them. Researchers were able to find two systems, which are supported by science and United States Patents. One of these devices is being marketed under Cell/Wave Guard. They found that up to 61 per cent of radio frequency emissions could be prevented from entering the body. While this represents a significant reduction it is not known if it is enough to guard against all potential effects. However, it is the best technology easily available today.

Detection Meter

The *Cell Sensor* measures the ELF 60 cycle fields as well as emissions from cellular phones and can be used to select low emission units. It can also be used to evaluate headsets (some headsets carry a signal like an antenna).

With the *Cell Sensor,* one can walk the house and in 5 minutes find these hot spots that measure over 2 and avoid placing beds and furniture there. Simple to use, the meter shows the hot spots with needle, light and sound. The meter is normally quiet.

When you walk by a hot spot, the needle goes up over 2, the light glows, the sound beeps! Without the meter, it is impossible to know where the hot spots are. A hot spot could be created by sleeping with a TV on the other side of the wall, or by faulty wiring, typical after renovation projects, or by proximity to visible or hidden underground powerlines!

While the reason for the creation of hot spots can be simple or complex, avoiding the hot spots is always very simple. Most hot spots are small, so moving a bed or furniture just a couple of feet usually does the trick! Most hot spots are dealt with that way, just avoid placing furniture there by moving it a little bit! Simply plug the included probe into the meter, turn it on, place the setting on H for high sensitivity, and the meter needle now is reading 0 to 5 in its display. Readings over 2 are high. Those are the hot spots to avoid. In this same way the meter shows safe distances from appliances such as TVs and microwaves!

Physical Effects

Our ability to generate the entire spectrum of electromagnetic fields is a double-edged sword. With no understanding of the relationship of these fields to living organisms, we have produced a global environmental alteration that has profound implications for human health. This has exposed all living organisms, from microorganisms to humans to novel energetic fields that never before existed. Complex electronic-resonance effects operate at the level of individual ions or molecules and can occur even in the absence of cells. If a charged particle or ion is exposed to a steady magnetic field, it will begin to go into a circular, or orbital motion at right angles to the applied magnetic field. The speed with which it orbits will be determined by the ratio between the charge and the mass of the particle and by the strength of the magnetic field. While the effects of complex electronic-resonance phenomena at the level of ions and molecules are significant, the resultant change in the function or structure of cells is far more important.

The effects of field exposure at the level of the total organism are the sum of the effects at the molecular level, the cellular level, and the level of the specific organs designed to be sensitive to the Earth's normal EMF environment. Field effects on entire functioning organisms are, therefore, a cascade of changes that finally result in many different structural, functional, and behavioral alterations, such as the new diseases that are emerging today. All abnormal, man-made electromagnetic fields, regardless of their frequencies, produce the same biological effects. Some of them are:

- Effects on growing cells, such as increases in the rate of cancer-cell division.

- Increases in the incidence of certain cancers
- Developmental abnormalities in embryos
- Alterations in neurochemicals, resulting in behavioral abnormalities such as suicide
- Alterations in biological cycles
- Stress responses in exposed animals that, if prolonged, lead to declines in immune-system efficiency
- Alterations in learning ability.

These bioeffects interact to weaken the health status of a human being who is chronically exposed to any abnormal field. The stress effect results in many stress-related diseases. If prolonged, this weakens the immune system, resulting in more inflammatory disease and cancer. The simultaneous effect of enhanced cancer-cell growth and the increase in the malignant changes of these cells leads to more cancer with faster-than normal rates of growth, especially in tissues with continuous rates of cell division.

The effect of field exposure on the human fetus is particularly important. Exposure to electromagnetic fields will result in the combined alterations of the speed and timing of fetal cell division, impairment of the genetic apparatus of dividing fetal cells, and the production of chromosomal abnormalities in the germ cells of the parents, all contribute to imbalance of normal development. Research has found a mechanism by which a very weak electrical signal can trigger a series of reactions that promote cancer by leading to unregulated cell growth.

The cell membrane has a huge electrical gradient across it. A very, very weak signal at some point along the membrane can trigger a tremendous increase in the amount of energy released at the surface of the cell, he says. The result could be a disruption in communication within and between cells causing them to begin growing out of control. This means that EMFs so weak that they were long believed not to have any physiological significance undoubtedly can influence the behaviour of cells through this amplification of energy mechanism. The cell membranes can detect a signal so weak that it would compare to your taking a 1.5 volt flashlight battery, putting a wire from one end of it in the Pacific Ocean at San Diego and putting a wire from the other end in Puget Sound to make a circuit including the ocean. Over that 1,200 mile wire, you'd get about one ten-millionth of a volt per centimeter. This is a very powerful mechanism for amplifying energy at the cell level. Exposure of human T-cell lymphocytes in culture to a low-strength 60-Hz electric field for forty-eight hours significantly reduces their cytotoxic ability against foreign cells.

New Kind of Illness

Reports from cities throughout the world indicate a new kind of illness that coincides in every case with the activation of a Personal Communication Services network. The symptoms are striking: pressure behind the eyes; dry, puffy lips; swollen thyroid; sudden rise in pulse rate and blood pressure; pressure or pain in the chest; insomnia; dizziness; headache; nausea; loss of appetite; coughing or wheezing; sinus problems; testicular or pelvic pain; muscle spasms; tremors; irritability; memory loss; pain in the legs or the soles of the feet; pains that move around the body; varying degrees of dehydration; and occasionally fever, rash or nosebleeds.

The illness appears to be confined to geographical areas served by new PCS and other digital systems. Despite well-documented exposes, the industry continues to deny that this radiation has any effect on life forms. Much as the asbestos and tobacco industries have done, the telecommunications industry has suppressed damaging evidence about its technology since at least 1927, it was discovered that red blood cells exposed to radio frequency waves (at levels far less powerful than permitted today) are forced to line up in chains resembling strings of pearls. A small group of scientists control the setting of health hazard standards, controlling what research bearing on that standard gets funded or published, while providing testimony for various companies and government agencies to the effect that substantial microwave energy exposure is safe. This small group of scientists is made up of engineers and veterinarians, not doctors, biologists, or epidemiologists. The American National Standards Institute (ANSI) was and still is setting microwave exposure standards. It is a private organization funded and controlled by industry.

AC ELECTRIC FIELDS

Electric fields are not very strong in most parts of a house. High electric-field areas are found near TVs, computer monitors (including laptops), fluorescent lights, light dimmer controls, and improperly grounded equipment. field strength drops off rapidly if at least 3 feet away. Electric fields are high near high-voltage power lines.

Polarization and Grounding

In household wiring, the wall outlet usually has a slot for a large prong and one for a small prong. This type of outlet is called polarized. The small slot is electrically hot; in a typical 115 volt system it oscillates 60 times a second between about +170 volts and –170 volts, compared to ground. To get a shock, it is necessary to touch this "hot" slot (or to touch something electrically connected to it), while also touching something that connects to ground. Things that are connected to ground include the larger neutral slot, the round grounding slot in a 3-prong wall outlet, plumbing pipes, and any metal that is part of a sink or bathtub. Besides being a shock hazard, an improperly grounded appliance produces AC electric fields, which creates currents in your body by proximity, even if you are not touching the appliance. You can sometmes detect improper grounding by lightly sliding a fingertip across the surface using very little pressure. You will usually feel a weak vibration if the surface is "hot." You can also use an AC voltmeter (connected between a known ground and the appliance case) or an AC electric field meter to determine if the case is hot.

Shielding Electric Fields

Shielding strong electrical fields can be done by using any type of conductive sheet material, such as aluminum window screen, in front of an indoor appliance. The shield should be connected to electrical ground for maximum efficiency.

AC Magnetic Fields

Magnetic fields are much more common in the home than are electric fields. They don't represent a shock hazard, but, like electric fields, they produce current in the body. Any wire that carries an AC electrical current produces magnetic fields.

Magnetic Field Sources

The main sources of AC magnetic fields in a home are transformers, motorized equipment, sloppy wiring inside the house walls, excess current carried by plumbing, and power lines or underground power cables.

Running cars have a strong AC magnetic field especially nearest the front floorboard, even theough the system is classified as DC. Commercial aircraft also have a strong field inside. The fields in cars and aircraft oscillate faster than the 60 times per second typical of household AC power. Wall outlets generally produce no magnetic fields. Any piece of electronic equipment that plugs into the wall, such as a clock radio or answering machine, will have a transformer.

Transformers and Motors

This transformer either plugs directly into the wall (AC adapter), or is built inside the equipment. The magnetic field is strong up to three feet from this transformer, whenever the power cord or AC adapter is plugged in. This field exists even if the appliance is not turned on, or even if the adapter is not connected.

If an electronic device uses an AC adapter, to avoid high field you should stay three feet away from the adapter, but you need not avoid the electronic device itself. However, if the device has an internal AC adapter and a plain power cord, stay three feet away from the device itself, whenever the cord is plugged in. Microwave ovens also have a large transformer built in, but they emit a magnetic field only while cooking. Ordinary incandescent light bulbs (and electric blankets made since 1992) produce very little magnetic field. However, high-intensity halogen desk and floor lamps have a large 12-volt transformer built into their bases and usually produce a strong field up to three feet away. Fluorescent lights use a smaller transformer and usually have little magnetic field past three feet so they are not a field source unless the light is used in close quarters. Motorized equipment also produces strong magnetic fields. Generally speaking, the higher the power, the stronger the field.

The field is low when you're at least three feet away from vacuum cleaners, motorized alarm clocks, and electric can openers, six feet from washing machines and furnace motors; and 18 inches form electric razors, hairdryers, and battery-powered motorized equipment.

Non-Localized Magnetic Fields

Non-localized magnetic fields are harder to avoid than the limited fields from appliances. Occasionally a nearby power line will produce this type of field. More often, household plumbing, carrying current, causes it. This situation is more likely to occur when the cold water pipe enters the opposite side of the building from the power line connection.

Safety codes require that the indoor plumbing be connected to the neutral wire of the power line. This can result in current flow.

A plumber can reduce magnetic fields which result from this situation by installing a non-conductive coupling on the cold water pipe just ouside the building. Non-localized fields due to household wiring are usually highest in the room next to where the power line connects or where there is a circuit breaker panel.

Detecting Magnetic Fields

A reliable way to detect the presence of strong fields is by holding a small magnet. When held lightly, it can be felt to vibrate when held in a 500 milligauss or higher field. Transformers and motors produce this much field a few inches away. A TV or computer monitor will start to jiggle or lose sharpness in a 30 milligauss or higher field, so this can be a more sensitive indicator of strong fields. A better indicator is a large coil of wire connected to an AC voltmeter. For the highest degree of accuracy, use an AC gaussmeter, specifically designed to detect and quantify magnetic fields.

Shielding Magnetic Fields

Shielding magnetic fields is more difficult than shielding electric fields. Sheets of galvanized mild steel work fairly well and are available in any hardware store. Use a sheet that is thin enough to cut with scissors, and note that two thin layers shield more than one thicker layer. Shielding should be placed so it is between you and the high-field source. Simply staying away from high-field areas is easier and more convenient.

Microwaves and Radiowaves

The strongest source of microwaves in a home is a microwave oven with a damaged door seal, or a cellular telephone. The new 2.2 gigahertz cordless phones also emit microwaves in the house. You should periodically check the microwave oven door to make sure it is not dented and that it latches shut fully parallel to the oven body. Correctly functioning microwave ovens emit somewhat less microwave power. Cellular phones and other radio sources sometimes produce interference, which can cause sensitive equipment, such as computers, to malfunction.

Shielding and Detecting Microwaves

Ordinary screen-door aluminum wire mesh (not plastic) can shield computers, etc. quite well if the screen is placed between the radio source and the computer. The screen should be electrically connected to ground. The more of the computer's area that's covered, the better the shielding. Also, the use of a power-conditioning outlet strip (specifying reduced electromagnetic interference) is recommended to avoid radio waves being picked up from the power line by the computer.

Chapter 4

Gravity and Electromagnetism

ELECTROMAGNETIC FIELD THEORY

It was the scientist Michael Faraday who first studied in detail the phenomena involving the interaction between electricity and magnetism. Amongst his many achievements, he is credited with the construction of the first electric motor and the discovery of both the principle and the method whereby a rotating magnet can be used to create an electric current in a coil of wire.He also observed the way in which iron filings arrange around a magnet. They appear to follow lines of magnetism leading out from one magnetic pole and back to the other pole. He suggested that the effect of a magnet on a wire carrying a current is the result of "lines of force". Faraday spent the latter part of his life working to devise an experiment that would confirm his theory of electromagnetic fields but without success.

Finally, a Scottish mathematician, James Clerk Maxwell, picked up on Faraday's ideas and was successful in developing a theory of electromagnetism which enabled testable predictions to be made. Inspired by Faraday's lines of force, he developed a model that unified magnetic and electrical forces. Maxwell "unified" the study of electricity and magnetism in four tidy equations. In essence he discovered that electric and magnetic fields were intrinsically related to one another, with or without the presence of a conductive path for electrons to flow.

'A changing electric field produces a perpendicular magnetic field' and 'A changing magnetic field produces a perpendicular electric field'Quite interestingly Maxwell predicted that electromagnetism would be propagated through space at a finite rate and was struck by the similarity between the predicted speed of electromagnetism and the speed of light. From this connection sprang the idea that light was an electric phenomenon and the subsequent discovery of radio waves.

Classical electromagnetic theory has been remarkably robust to modern innovations. It was the inspiration for special relativity and did not need alteration. As quantum mechanics evolved, the field equations remained the same. One technical change was in the action, a gauge had to be chosen. The advent of general relativity did not alter electromagnetism. Instead, general relativity provided the background metric needed by electromagnetism. The U(1) symmetry of electromagnetism served as the starting point for the formation of the standard model, which unifies electromagnetism, the weak, and the strong forces. A problem with this picture is the separation of gravity from the forces of the standard model. In this paper, I use electromagnetic theory to deduce the laws that govern gravity.

THE ACTION

The classical action for electromagnetism has three terms, a contraction of the irreducible anti symmetric field strength tensor and a coupling of electric charge with the potential, and an inertia term. To be logically complete, Nature must also exploit a contraction of an irreducible symmetric field strength tensor. The symmetric field will be associated with a different charge current that couples with the potential because the tensor is irreducible and thus an independent fundamental force. This new force must be long range like electromagnetism. The only such force is gravity.

Like charges repel in the electromagnetic action for several reasons. First, the signs of the charge coupling and the field strength tensor contraction are the same. The field equations that arise from varying the action with respect to the potential keeping the velocity fixed will have like charges repel. Second, the signs of the charge coupling term and the inertia term are the same. Varying the action with respect to velocity keeping the potential fixed generates the Lorentz force where like charges repel. Third, the field strength tensor is anti symmetric, so it must be represented by an odd spin particle, the spin 1 photon. For gravity, like charges attract. The coupling term must have the sign opposite to the field strength contraction and inertia terms so both the field and force equations will have like charges attract. A traceless symmetric rank 2 tensor can be represented by a spin 2 particle that travels at the speed of light like the photon, so the particle matches the description of a graviton. Based on these deductions, the action for gravity and electromagnetism (GEM) can be written:

$$_{GEM} = \int \sqrt{-gd^4}$$

$$x\left(-\frac{pm}{\gamma}-\frac{1}{c}J^{\varphi}\left(A_{\varphi}+I_{\|}A_{\varphi}I_{\|}\right)+\frac{1}{c}J^{\varphi}\left(I_{tr}\left(A_{\varphi}-I_{\|}A_{\varphi i\|}\right)I_{tr}\right)\right)^{+}_{+}$$

$$-\frac{1}{4C^2}(\nabla_{\varphi}A_{v}-\nabla_{v}A_{\varphi})(\nabla^{\varphi}A^{v}-\nabla^{v}A^{\varphi})-$$

$$\frac{1}{4c^2}\left((\nabla_{\varphi}A_{v}+\nabla_{v}A_{\varphi})(\nabla^{\varphi}A^{v}+\nabla^{v}A^{\varphi})\right)$$

The current coupling term appears more complex than one might anticipate. The reason is that this current coupling must represent both a spin 1 and a spin 2 current-current interaction. One is free to choose the orientation of the theoretical 4-potential with the actual 4-current, derived from the 4-potential via a Fourier transformation. Where there is freedom, there is a physics message. In this case, we have freedom with the phase for the current-current interaction. The part of the potential that is tranverse to the current can be separated from that which is parallel. The phase of the current-current interaction for the transverse potential-current has the symmetry of a spin 1 field, as is needed to have like charges repel. Rotating the longitudinal current around the parallel axis accomplishes two things. First, the sign of the coupling term must change to conserver the correct signs of the contraction of the current density with the potential. Second, the phase of the current-current interaction for the rotated longitudinal potential-current has the symmetry of a spin 2 field. A even spin field is required for like charges to attract.

The current density is made of difference between the electric current density and the mass current density, using the square root of Newton's gravitational constant with the mass current density so the units are identical. Electric charge is only known to ten significant digits, while for fundamental particles, the mass in electrical units can be more than thirteen orders of magnitude smaller. At this time, I have no notion why there is this large difference, but it justifies the practice of working on problems in electromagnetism separately from gravity.

The deep insight into gravity provided by general relativity is that the geometry of spacetime is not presumed. In special relativity, the metric must be constant, the same for all inertial observers. In general relativity, the metric can vary depending on where the observer is in spacetime: closer to a mass source will have more spacetime curvature. The Riemann curvature tensor is a rank 4 tensor that contains second order derivatives of the metric. The Hilbert action has the Ricci scalar, a contraction of the Riemann curvature tensor. By varying the Hilbert action with respect to the metric tensor, the geometry of spacetime can change as dictated by the Einstein field equations.

Electromagnetism works no matter what the metric happens to be. The antisymetric tensor cannot characterize how the symmetric metric tensor changes, so the possibly dynamic metric must be supplied as part of the background mathematical structure. By including a symmetric field strength tensor in the GEM action, it becomes possible for the action to account for the changes in metric tensor, and thus remove the metric from the background structure. The idea is not to treat the metric as an active field, but instead to use a symmetry of the action to provide a constraint on a metric that varies in spacetime.

The way gravity in the GEM action works mathematically is different from general relativity. The Hilbert action requires only one field, the metric field, to be varied in the action to generate the Einstein field equations. The GEM action does not treat the metric as a field. Instead there is a diffeomorphism symmetry of the action that allows one to chose how much of a covariant derivative is due to a dynamic potential or a dynamic metric. Where there is a symmetry, there is a conserved charge. Since the symmetry involves the metric, the charge must be mass. Energy and momentum also arise from a symmetry of the action. Since mass is the square of the energy minus the square of the momentum, it is logically consistent that energy, momentum, and mass all can be viewed as different symmetries of the same action.

The action has symmetry for mass less particles. Mass charge breaks symmetry. This is an attractive feature because it means the Higgs mechanism for introducing mass into the standard model is unnecessary. The GEM proposal predicts the Higgs boson will not be found in particle field experiments such as the Large Hadron Collider. The trace of the symmetric field strength tensor is a scalar field which plays the role of the Higgs.

FIELD EQUATIONS AND SOLUTIONS

The field equations can be determined in the usual way by fixing the metric tensor up to a diffeomorphism and then varying the action with respect to the potential:

$$J^{\mu}_{tr} - J^{\mu}_{\|} = \nabla_{\mu}\nabla^{\mu} A^{\mu}$$

This 4D wave equation has the ability to describe 4 modes of emission of particles that travel at the speed of light. The two transverse modes are electromagnetism. That leaves the scalar and longitudinal modes for the graviton.

It is critical to note in equation. wave the scalar D'Alembertian operator is not acting alone on the potential. Instead, two covariant derivatives are applied to the potential.

A covariant derivative contains both the normal derivative and the connection. The complete wave equation is composed of the D'Alembertian operator acting on the potential, the divergence of the connection, the connection of the first derivative of the potential, and the connection of the connection.

Spacetime is nearly flat if one excludes regions near black holes, so the connection of the connection term can be ignored. The divergence of the connection contains second derivatives of the metric, so in principle could provide an equation that must be solved to determine a smoothly changing metric. Because equation wave is valid up to a diffeomorphism of the metric, the equation cannot be solved until that degree of freedom is eliminated. In this paper, I will work with metric compatible, torsion-free connections as is done in general relativity.

If one chooses to work with a flat Minkowski metric in Euclidean spacetime, the connection will be zero everywhere. For a static point charge, the first field equation is a sum of the static form of Gauss' law of electric charge and Newton's law of gravity:

$$\rho_{tr} - \rho_{\|} = -c\nabla^2\phi$$

The repulsive force of like electric charges is decreased by mass attraction. This field equation can account for all classical effects of Gauss' and Newton's law because it is identical to their superposition. There are two articles in the literature that assert without proof that one cannot form vector field equations where like charges attract. Perhaps there was an aversion to putting a negative sign in front of the mass. What matters are the relative signs, that the mass charge density has the same sign as the Laplacian operator in equation.

A flaw of Newton's law of gravity is that it requires changes in the charge density to propagate instantaneously, which is not consistent with special relativity. One road to Einstein's field equations is to make Newton's law respect special relativity. Here is the dynamic form of the equation

$$\rho_{tr} - \rho_{\|} = \frac{1}{c}\frac{\partial^2 \phi}{\partial t^2} - c\nabla^2 \phi$$

The waves for gravity and light both propagate at the speed of light. This equation transforms as the first term of a 4-vector equation and so is manifestly covariant under a Lorentz transformation.

The gauge choice-how one measures things-for the GEM action concerns how to measure a covariant derivative: how much should be due to the standard derivative versus how much is due to the connection, the changes in the metric? In the preceding paragraphs, a choice was made that everything was due to the standard derivative of the potential. This time I choose to allow all of the change be due to the connection, nothing due to the standard derivative of the potential. The first field equation under this gauge choice for a static charge is:

$$\rho_{tr} - \rho_{\|} = -\partial_{\mu}\Gamma^{0\mu}_{\mu} A^{\mu}$$

A metric must be found whose connection solves this first-order partial differential equation. This might appear like a difficult task since the connection has three derivatives of the metric. Yet one of the derivatives is zero because the charges are static. If the metric were diagonal, another of the derivatives would be zero. The metric must reduce to the Minkowski metric as the charges go to zero. For an electrically neutral source, the metric becomes the Rosen metric. That metric is not a solution to the Einstein field equations, but has been studied in the literature because it is consistent with experimental tests of gravity to first order parameterized post-Newtonian (PPN) accuracy. At second order PPN accuracy, equation exponential metric, predicts 0.7 microarcseconds more bending of star light by the Sun. An experiment such as LATOR, could confirm or reject this proposal on purely experimental grounds.

The bimetric theory of Rosen is not consistent with strong field tests of gravity. The fixed background metric can store momentum, allowing for bipolar modes of emission of gravity waves which are ruled out by observations of binary pulsars.

The GEM action does not have an additional field to store momentum. For an isolated source, the lowest mode of emission that conserves momentum will be the quadrapole mode, consistent with observation.

QUANTIZATION

The 4D wave equation, has already been quantized. They were quantizing the field only to do the work of electromagnetism in a manifestly covariant way. The transverse modes of emission do the work of electromagnetism. The scalar mode of a spin 1 field caused a serious technical problem: it suggested negative probability states. To solve this issue, a supplementary condition was imposed to eliminate both the scalar and longitudinal modes.

There are two independent ways to spot the spin 1 field in the classical electrodynamic Lagrangian. The first is by looking at the anti-symmetric, rank two field strength tensor, $\nabla_{\mu} A_{\mu} - \nabla_{\mu} A_{\mu}$. If the indexes are swapped, then all the signs flip, the sign of an odd spin tensor.

The second approach considers current-current interactions of the charge coupling term,-$J^{\mu} A_{\mu}$. One can take the Fourier transformation of the 4-potential and write it as a current density in the momentum representation,-$(1/k^2)J' J^{\mu}$. If one constrains the current to move along the z axis, conservation of electric charge can be used to eliminate the current density. The current-current interaction is thus:

$$-\frac{1}{k^2} J_q^{'\mu} J_{\mu} = -\frac{1}{w^2} j_{\mu}^{'} j_{\mu} + \frac{1}{w^2 - k^2}\left(j_x^{'} j_x + j_y^{'} j_y\right)$$

The current along the z axis is real. The currents along x and y are for the two independent polarizations of virtual photons needed for relativistic corrections. What is the symmetry of these virtual photons? Form the product of these two virtual currents:

$$\left(0, j_x^{'}, j_y^{'}, 0\right)\left(0, j_x, j_y, 0\right) = \left(j_x^{'} j_z + j_y^{'} j_y, 0, 0, j_y^{'} j_z - j_z^{'} j_y\right)$$

The phase part of this product (the z component) will require a 2π rotation to return its original location, a defining characteristic of a spin 1 field. The GEM action has the antisymmetric field strength tensor, and this kind of current-current interaction. On a quantum level, the work of photons can be done.

The GEM action also has a spin 2 field. There is a symmetric, rank 2 field strength tensor, $\nabla_{\mu}A_{\mu} - \nabla_{\mu}A_{\mu}$. If the indexes are swapped, then all the signs stay the same, the sign of an even spin tensor. Since the tensor has indexes, it cannot be a scalar field. A spin 0 field can be formed by taking the trace of the symmetric field strength tensor, a Lorentz invariant quantity. If the trace is zero, then the particle can travel at the speed of light c. If the trace is not zero, then the scalar field formed will break gauge symmetry. In other words, a scalar inertial mass field breaks gauge symmetry. The Higgs mechanism is not necessary.

The analysis of the current-current density proceeds along lines similar to the case for electromagnetism. There is a difference for calculating the product of two currents. The real current is the one along z, yet the current along z does not appear in Equation. Although the reader may realise the currents along x and y in Equation are virtual, it would be preferable to enforce that they are zero. Working with pairs of currents with xz and yz, it may be possible to do both. The conjugate used in Equation is not the only anti-involutive automorphism. There is also a conjugate that flip all the signs but the first term of the 3-vector referred here as the first conjugate. One can also construct the second conjugate to keep the sign of the y current positive while flipping all others.

These products have the character of spin 2 because the two parts add together, allowing a return after only a ð rotation. If the three product currents are added together, The result has the real current along the z axis, and phases for spin 1 and spin 2 polarization modes. The four modes account for the four degrees of freedom in the 4-current. The scalar and longitudinal modes of emission will do the work of gravity. General relativity predicts transverse modes of emission for gravity waves. Should a gravitational wave detection experiment succeed, the polarization of the gravity wave will serve as an experimental test of this proposal. There is a strong belief that field equations for gravity must be nonlinear. To be more precise, this means that gravity fields gravitate. The same is not true for electromagnetism: the electromagnetic field cannot be a source of charge. There are thought experiments to support the notion that gravity must be nonlinear.An example from a review paper on general relativity had two boxes with six neutral particles in each. Imagine that the rest mass of one particle were to be converted entirely into the kinetic energy of the other five. Would the first box be able to tell any difference in energy density and spacetime curvature between the two? If the answer is no, gravity must be governed by a nonlinear field equation.

Repeat the thought experiment, but this time make all the particles positively charged. Now the thought experiment cannot be done because it would involve destroying an electric charge. There are solid theoretical and experimental tests demonstrating that electric charge cannot be destroyed. Because the thought experiment cannot be done, the conclusion is not supported.

A NEW IMPLEMENTATION

The experimental tests of the equivalence principle indicate that gravity must be a metric theory. This paper questions how to implement a metric theory. General relativity takes a direct approach: start with the Riemann curvature tensor that characterizes how a metric changes, and put its contraction the Ricci scalar into the action.

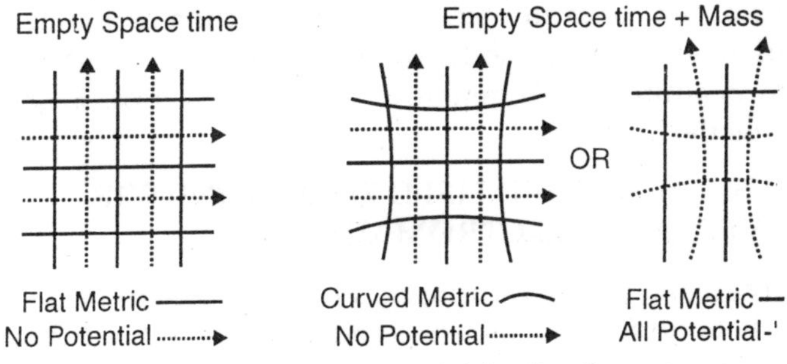

Fig. Metric or Potential for Gravity

This paper uses the symmetry found in the definition of a covariant derivative to allow spacetime geometry to change based on what mass is in that spacetime. It is the flexibility of the spacetime geometry that indicates the approach can be characterized as a metric theory.

There is a well-known thought experiment of a man in a closed box who would be unable to tell if he was in a smoothly accelerating rocket ship or the box was sitting on the surface of a planet ignoring tidal effects.

For the box on the planet, general relativity would be able to generate a metric equation that in a limit process, would become a potential theory. For the GEM approach, one could choose a dynamic metric which would be equivalent to first order PPN accuracy to the metric from general relativity, or with the practical power of a diffeomorphism, choose to work in a flat spacetime and have a 4-potential completely characterize the gravitational field. A scalar potential like Newton's law of gravity is inadequate because there is only one parameter, so it gets the answer to light bending around the Sun half right. With four parameters, a 4-potential can describe the way measurements of both time and space are bent by a 4-momentum current density.

The GEM action is to a surprising degree exactly what Einstein searched for over the last half of his life. He tried an impressive range of ideas, from five dimensional Kaluza-Klein variations, to asymmetric Riemann curvature tensors, all to no avail. The radical approach used here is to abandon the Riemann curvature tensor, and work only with the building blocks of the curvature tensor, the connection as found in a covariant derivative. The 4D wave equation in this paper was called beautiful by Feynman in his lecture series. Elegance governs the heavens.

CLASSICAL UNIFIED FIELD EQUATIONS FOR GRAVITY AND ELECTROMAGNETISM

The equations governing gravity and electromagnetism show both profound similarities and unambiguous differences. Albert Einstein worked to unify gravity and electromagnetism, mainly by trying to generalize Riemannian geometry. Hamilton's quaternions are a 4-dimensional topological algebraic field related to the real and complex numbers equipped with a static Euclidean 4-basis. Riemannian quaternions as defined herein explicitly allow for dynamic changes in the basis vectors.

The equivalence principle of general relativity which applies only to mass is generalized because for any Riemannian quaternion differential equation, the chain rule means a change could be caused by the potential and/or the basis vectors. The Maxwell equations are generated using a quaternion potential and operators. Unfortunately, the algebra is complicated. The unified force field proposed is modeled on a simplification of the electromagnetic field strength tensor, being formed by a quaternion differential operator acting on a potential.

This generates an even, antisymmetric-matrix force field for electricity and an odd, antisymmetric-matrix force field for magnetism, where the even field conserves its sign if the order of the differential and the potential are reversed unlike the odd field. Gauge symmetry is broken for massive particles by the even, symmetric-matrix term, which is interpreted as being due to gravity. In tensor analysis, a differential operator acting on the field strength tensor creates the Maxwell equations.

The unified field equations for an isolated source are generated by acting on the unified force field with an additional differential operator. This contains a quaternion representation of the Maxwell equations, a classical link to the quantum Aharonov-Bohm effect, and dynamic field equations for gravity. Vacuum and zero net current solutions to the unified field equations are discussed.

The field equations conserve both electric charge density and mass density. Under a Lorentz transformation, the gravitational and electromagnetic fields are Lorentz invariant and Lorentz covariant respectively, but there are residual terms whose meaning is not clear presently. An additional constraint is required for gauge transformations of a massive field.

Einstein's Vision using Quaternions

Three of the four known forces in physics have been unified via the standard model:the electromagnetic, the weak, and the strong forces. The holdout remains gravity, the first force characterized mathematically by Isaac Newton. The parallels between gravity and electromagnetism are evident. Newton's law of gravity and Coulomb's law are inverse square laws. Both forces can be attractive, but Coulomb's law can also be a repulsive force. A longstanding goal of modern physics is to explain the similarities and differences between gravity and electromagnetism.

Albert Einstein had a specific idea for how to formulate an acceptable unified field theory taken from A. Pais,-"Subtle is the Lord" the science and life of Albert Einstein". One unusual aspect of Einstein's view was that he believed the unified field would lead to a new foundation for quantum mechanics, an idea which is not shared by some of today's thinkers. Most of Einstein's efforts over 40 years were directed in a search to generalize Riemannian differential geometry in four dimensions.

To a degree which has pleasantly surprised the author, Einstein's vision to unify gravity and electromag-netism has been followed. The construction of a new 4-dimensional geometry is dictated by insights garnered from physics. Events in spacetime are composed of a scalar for time and a 3-vector for space. The four-dimensional topological algebraic field of quaternions has the same structure, so quaternions will be the starting point of this effort.

Laws of physics are expressed in a coordinate-independent way. The sum or difference of two quaternions can only be defined if the two quaternions in question share the same 4-basis. Riemannian quaternions make coordinate-independence explicit. In special relativity, regions in spacetime are delimited by the light cone, where the net change in 3-space is equal to the net change in time. The parity between changes in 3-space and time is constructed into the definition of a Riemannian quaternion. In general relativity, the field equations make the metric a dynamic variable. The basis vectors of Riemannian quaternions can be dynamic, so the metric can be dynamic. The dynamic nature of the basis vectors leads to the general equivalence principle, whereby any law, even those in electromagnetism, can be the result of a change in reference frame.

Physical laws are the result of simple Riemannian quaternion differential equations. First-order Riemannian quaternion differential equations create force fields for gravity, electricity, and magnetism. Second-order differential equations create dynamic field equations for gravity, the Maxwell equations for electromagnetism, and a classical counterpart to the Aharanov-Bohm effect of quantum mechanics. Third-order differential equations create conservation laws. Homogeneous solutions to the second order differential equations are related to gauge symmetry.

The second paper in this series of three investigates a unified force law, with a focus on a particular solution which may eliminate the need for dark matter to explain the mass distribution and velocity profile for spiral galaxies. The third paper develops a new approach to quaternion analysis. The equations of the first two papers are recast with the new definition of a quaternion derivative, resulting in a quantum unified field and force theory.

Events in Spacetime and Quaternions

An event in spacetime is considered by the author as the fundamental form of information in physics. Events have structure. There are four degrees of freedom divided into two dissimilar parts:time is a scalar, and space is a 3-vector. This structure should be reflected in all the mathematics used to describe patterns of events. For this reason, this paper focuses exclusively on quaternions, the 4-dimensional number where the terms scalar and vector where first used.

Hamilton's quaternions, along with the far better know real and complex numbers, can be added, subtracted, multiplied and divided. Technically, these three numbers are the only finite-dimensional, associative, topological, algebraic fields, up to an isomorphism.

Number	Dimensions	Totally Ordered	Commutative
Real	1	Yes	Yes
Complex	2	No	Yese
Quaternions	4	No	No

Hamilton's quaternions have a Euclidean 4-basis composed of 1, i, j, and k. The rules of multiplication were inspired by those for complex numbers: $1^2=1$, $i^2=j^2=k^2=ijk=-1$. Quaternions also have a real 4 × 4 matrix representation:

$$q\left(t, x, y, z\right)=\begin{pmatrix} t & -x & -y & -z \\ x & t & -z & y \\ y & z & t & -x \\ z & -y & x & t \end{pmatrix}$$

Although written in Cartesian coordinates, quaternions can be written in any linearly-independent 4-basis because matrix algebra provides the necessary techniques for changing the basis. Therefore, like tensors, a quaternion equation is independent of the chosen basis. One could view quaternions as tensors restricted to a 4-dimensional algebraic field. For the sake of consistency, all transformations are also constrained to the same division algebra. This constraint might first appear too restrictive since for example it eliminates simple matrices for row permutations. Since quaternions are an algebraic field, there necessarily exists a combination of quaternions that achieves the action of a permutation. The need for consistency will overrule convenience.

Laws in physics are independent of coordinate systems. To make the coordinate independence explicit, amplitudes and basis vectors will be separated using a new notation. Consider a quaternion 4-function, $A_n=(a_0, a_1, a_2, a_3)$, and an arbitrary 4-basis, $Ihat_n = (ihat_0, ihat_1, ihat_2, ihat_3)$. In spacetime, the line that divides causality is define by the light cone. On the light cone, the total change in 3-space over the change in time is equal to one. Physics therefore indicates parity between the total 3-vector and the scalar, instead of weighing all four equally. The scaling factor of a third for the 3-vector plays a vital role in the definition of a regular function in the third paper of this series.

The equivalence principle of general relativity asserts, with experiments to back it up, that the inertial mass equals the gravitational mass. An accelerated reference frame can be indistinguishable from the effect of a mass density. No corresponding principle applies to electromagnetism, which depends only on the electromagnetic field tensor built from the potential. With Riemannian quaternions, the 4-unit vector doe s not have to be static by taking the time derivative of the first term and using the chain rule:

$$\frac{\partial a_0 \hat{i}}{\partial i_0} = \hat{i}_0 \frac{\partial a_0}{\partial i_0} + a_0 \frac{\partial \hat{i}_0}{\partial i_0}$$

The unit vector for time, ihat0, can change over an infinitely small amount of time, i0. Any change in a quaternion potential function could be due to contributions from a change in potential, the ihat0 da0/di0 term, and/or a change in the basis, the a0 dihat0/di0 term. Is this mathematical property related to physics? Consider Gauss' law written with Riemannian quaternions:

$$-\frac{\hat{i}_n^{\,2}}{9} \frac{\partial E_n}{\partial i_n} - \frac{\hat{i}_n E_n}{9} \frac{\partial i_n}{\partial i_n} = 4\pi\rho, \quad n = 1,2,3$$

The divergence of the electric field might equal the source, or equivalently, the divergence of the basis vectors. The "general equivalence principle" as defined here means that any measurement can be due to a change in the potential and/or a change in the basis vectors. The general equivalence principle is applicable to both gravity and electromagnetism.

Metrics and Quaternion Products

The theories of special and general relativity dictate the distance between events in spacetime. Although fundamentally different in their mathematical structure, inertia is a link between the two. Special relativity dictates the transformation rules for observers who change their inertia, assuming the system observed does not change. The field equations of general relativity detail the changes in distance due to a system changing its inertia from the vacuum to a non-zero energy density. A quaternion product necessarily contains information about the metric, but also has information in the 3-vector. This additional information about quaternion products will suggest a provocative link between metrics and inertia consistent with both special and general relativity.

Most structures in Nature do not transform like a scalar and a 3-vector. Quaternion products multiply two 4-basis vectors, and those products will transform differently. The rules of quaternion multiplication mirror those of complex numbers. Instead of the imaginary number i, there is a unit 3-vector for each quaternion playing an analogous role. The difference is that unit 3-vectors do not all have to point in the same direction. Based on the angle between them, two different unit 3-vectors have both a dot and cross product. The dot and cross products completely characterize the relationship between the two unit vectors. Compare the product of multiplying two complex numbers (a, bi) and (c, di): (a, bi) (c, di)=(ac–bd, ad+bc), with two quaternions;

$$\left(a, B^{\bar{I}}\right) \text{ and } \left(c, D^{\bar{I}}\right),$$

Complex numbers commute because they do not have a cross product in the result. If the order of quaternion multiplication is reversed, then only the cross product would change its sign. Quaternion multiplication does not commute due to the behaviour of the cross product. If the cross product is zero, then quaternion multiplication has all of the properties of complex numbers. If, on the other hand, the only value of a quaternion product is equal to the cross product, then multiplication is anti-commutative. Individually, the mathematical properties of commuting and anti-commuting algebras are well known.

A quaternion product is the superp-osition of these two types of algebras that forms a division algebra. Several steps are required to square of the difference of two Riemannian quaternions to form a measure of distance. First, the basis of the two quaternions must be shared. It makes no sense to subtract something in spherical coordinates from something in Cartesian coordinates. The basis does not have to be constant, only shared. Every quaternion commutes with itself, so the cross product is zero. There are seven unique pairs of basis vectors in a square:

$$\left(da_0\,\hat{i}_0, dA_n\frac{\hat{I}_n}{3}\right)^2 = \left(da_0^2\,\hat{i}_0^{\,2}\frac{\hat{I}_n^{\,2}}{9}, 2\,da_0\,dA_n\frac{\hat{I}_0\hat{I}_n}{3}\right)$$

The signs were chosen to be consistent with Hamilton's quaternion algebra. The four square basis vectors i_mu^2 define the metric. If the basis vectors are not constant, then the metric is dynamic. Define a "3-rope" to be the three other terms, which have the form i0 In. Notice that the 3-rope starts in one time-space location and will have a non-zero length if it ends up at a different location and time. With quaternion products, the 3-rope is a natural companion to a metric for information about distance.

In special relativity, if the inertia of the observer but not the system is changed, the metric is invariant. The 3-rope is covariant, because it is known how it changes. Given the utility of duality, a complementary hypothesis to the invariant metric of special relativity would propose the if the inertia of the system but not the observer is changed, there exists a choice of basis vectors such that the 3-rope is invariant but the metric changes in a known way. If the magnitude of the time and 3-space basis vectors are inversely related, the magnitude of the product of the time basis vector with each 3-space basis vector will be constant even if the basis vectors themselves are dynamic. This hypothesis asserts the existence of such a basis, but that particular basis does not have to be used. Hamilton had the freedom to use the rule found in the above equation, but made the more obvious choice of $\left(\hat{i}_0\right)^2 = \left(-\hat{I}_n\right)^2$. The existence of a basis where the 3-rope is constant despite a change in the inertia of the system will have to be treated as provisional in this paper. In the second paper of this series, a metric with this property will be found and discussed.

Physically Relevant Differential Equations

Is there a rational way to construct physically relevant quaternion equations? The method used here will be to mimic the tensor equations of electromagnetism. The electromagnetic field strength tensor is formed by a differential operator acting on a potential. The Maxwell equations are formed by acting on the field with another differential operator. The Lorentz 4-force is created by the product of a electric charge, the electromagnetic field strength tensor, and a 4-velocity. This pattern will be repeated starting from an asymmetric field to create the same field and force equations using quaternion differentials and potentials. The challenge in this exercise is in the interpretation.

As a first step to constructing differential equations, examine how the differential operator (d/dt, Del) acts on a potential function (phi, A):

$$\left(\frac{\partial}{\partial t}, \overline{\nabla}\right)(\phi, \overline{A}) = \left(\frac{\partial \phi}{\partial t} - \overline{\nabla}.\overline{A}, \frac{\partial \overline{A}}{\partial t} + \overline{\nabla}\phi + \overline{\nabla} \times \overline{A}\right)$$

For the sake of clarity, the notation introduced for Riemann quaternions has been suppress, so the reader is encouraged to recognize that there are also a parallel set of terms for changes in the basis vectors. The previous equation is a complete assessment of the change in the 4-dimensional potential/basis, involving two time derivatives, the divergence, the gradient and the curl all in one. A unified field theory should account for all conceivable forms of change in a 4-dimensional potential/basis, as is the case here.

Quaternion operators and potentials have not been used to express the Maxwell equations. The reason can be found in the previous equation, where the sign of the divergence of A is opposite of the curl of A. In the Maxwell equations, the divergence and the curl involving the electric and magnetic field are all positive. Many others, even in Maxwell's time, have used complex-valued quaternions for the task because the extra imaginary number can be used to get the signs correct.

However, complex-valued quaternions are not an algebraic field. The norm, $t^2+x^2+y^2+z^2$, for a non-zero quaternion could equal zero if the values of t, x, y, and z were complex. This paper involves the constraint of working exclusively with 4-dimensional algebraic fields. Therefore, no matter how salutary the work with complex-valued quaternions, it is not relevant to this paper. The reason to hope for unification using quaternions can be found in an analysis of symmetry provided by Albert Einstein.

The "four-dimensional continuum" could be viewed as a technical constraint involving topology. Fortunately, quaternions do have a topological structure since they have a norm. Nature is asymmetric, containing both a symmetric metric for gravity and an antisymmetric tensor for electromagnetism. With this in mind, rewrite out the real 4x4 matrix representation of a quaternion:

$$q\left(t, x, y, z\right) = \begin{pmatrix} t & 0 & 0 & 0 \\ 0 & t & 0 & 0 \\ 0 & 0 & t & 0 \\ 0 & 0 & 0 & t \end{pmatrix} + \begin{pmatrix} 0 & -x & -y & -z \\ x & 0 & -z & y \\ y & z & 0 & -x \\ z & -y & x & 0 \end{pmatrix}$$

The scalar component can be represented by a symmetric 4 × 4 matrix, invariant under transposition and conjugation (these are the same operations for quaternions). The 3-vector component (x, y and z in the representation above) is off-diagonal and can be represented by an anti symmetric 4x4 matrix, because taking the transpose will flip the signs of the 3-vector. Quaternions are asymmetric in their matrix representation, a property which is critical to using them for unifying gravity and electromagnetism.

Recreating the Maxwell Equations

Maxwell speculated that his set of equations might be expressed with quaternions someday. The divergence, gradient, and curl were initially developed by Hamilton during his investigation of quaternions. For the sake of logical consistency, any system of differential equations, such as the Maxwell equations, that depends on these tools must have a quaternion representation.

The Maxwell equations are gauge invariant. How can this property be built into a quaternion expression? Consider a common gauge such as the Lorenz gauge, dphi/dt + div A = 0. In quaternion parlance, this is a quaternion-scalar formed from a differential quaternion acting on a potential. To be invariant under an arbitrary gauge transformation, the quaternion-scalar must be set to zero. This can be done with the vector operator, (q–q*)/2. Search for a combination of quaternion operators and potentials that generate the Maxwell equations:

$$\frac{\left(\square^{*}\mathrm{Vector}\left(\square^{*}A^{*}\right)-\square\mathrm{Vector}\left(\square\bar{A}\right)\right)}{2}=$$

$$=\left(\nabla-\left(-\frac{\partial\bar{A}}{\partial t}-\bar{\nabla}\phi\right),\frac{\partial\bar{\nabla}\times\bar{A}}{\partial t}+\bar{\nabla}\times\left(-\frac{\partial\bar{A}}{\partial t}-\bar{\nabla}\phi\right)\right)=$$

$$=\left(\bar{\nabla}-\bar{E},\frac{\partial\bar{B}}{\partial t}+\bar{\nabla}\times\bar{E}\right)=$$

$$=\left(4\pi\rho,\bar{0}\right).$$

This is Ampere's law and the no monopoles vector identity assuming a simply-connected topology. Any choice of gauge will not make a contribution due to the vector operator. If the vector operator was not used, then the gradient of the symmetric-matrix force field would be linked to the electromagnetic source equation, Ampere's law.

Generate the other two Maxwell equations:

This is Gauss' and Faraday's law. Again, if the vector operator had not been used, the time derivative of the symmetric-matrix force field would be associated with the electromagnetic source equation, Gauss' law. To specify the Maxwell equations completely, two quaternion equations are required, just like the 4-vector approach.

Although successful, the quaternion expression is unappealing for reasons of simplicity, consistency and completeness. A complicated collection of sums or differences of differential operators acting on potentials-along with their conjugates-is required. There is no obvious reason this combination of terms should be central to the nature of light. One motivation for the search for a unified potential field involves simplifying the above expressions.

When a quaternion differential acts on a function, the

divergence always has a sign opposite the curl. The opposite situation applies to the Maxwell equations. Of course the signs of the Maxwell equations cannot be changed. However, it may be worth the effort to explore equations with sign conventions consistent with the quaternion algebra, where the operators for divergence and curl were conceived.

Information about the change in the potential is explicitly discarded by the vector operator. Justification comes from the plea for gauge symmetry, essential for the Maxwell equations. The Maxwell equations apply to massless particles. Gauge symmetry is broken for massive fields. More information about the potential might be used in unification of electromagnetism with gravity. A gauge is also matrix symmetric, so it could provide a complete picture concerning symmetry.

One Unified force Field from one Potential Field

For massless particles, the Maxwell equations are sufficient to explain classical and quantum electrodynamic phenomena in a gauge-invariant way. To unify electromagnetism with gravity, the gauge symmetry must be broken, opening the door to massive particles. Because of the constraints imposed by quaternion algebra, there is little freedom to choose the gauge with a simple quaternion expression. In the standard approach to the electromagnetic field, a differential 4-vector acts on a 4-vector potential in such a way as to create an antisymmetric second-rank tensor. The unified field hypothesis proposed involves a quaternion differential operator acting on a quaternion potential:

$$\square^* A^* \left(\frac{\partial \phi}{\partial t} - \overline{\nabla}.\,\overline{A}, -\frac{\partial \overline{A}}{\partial t} - \overline{\nabla}\phi + \overline{\nabla} \times \overline{A} \right)$$

This is a natural suggestion with this algebra. The antisymmetric-matrix component of the unified field has the same elements as the standard electromagnetic field tensor. Define the electric field E as the even terms, the ones that will not change signs if the order of the differential operator and the potential are reversed. The magnetic field B is the curl of A, the odd term. The justification for proposing the unified force field hypothesis rests on the presence of the electric and magnetic fields.

In some ways, the above equation looks just like the old idea of combining a scalar gauge field with the electromagnetic field strength tensor, as Gupta did in 1950 in order to quantize the Maxwell equations. He concluded that although useful because it is written in manifestly relativistic form, no new results beyond the Maxwell equation are obtained. Examine just the gauge contribution to the Lagrangian for this unified field:

$$L = -\frac{1}{2}\left(\frac{\partial \phi}{\partial t}, -\frac{1}{3}\frac{\partial Ax}{\partial x}, -\frac{1}{3}\frac{\partial A_y}{\partial y}, -\frac{1}{3}\frac{\partial A_z}{\partial z}\right)^2$$

Take the derivative of the Lagrangian with respect to the gauge variables:

$$\frac{\partial L}{\partial \frac{\partial A_\mu}{\partial x_\mu}} = 0$$

By Noether's theorem, this conserved current indicates a symmetry of the Lagrangian. This is why the proposal involves new physics. The gauge is a dynamic variable constrained by the Lagrangian.

A quaternion potential function has four degrees of freedom represented by the scalar function phi and the 3-vector function A. Acting on this with one differential operators does not change the degrees of freedom. Instead, the tangent spaces of the potential will offer more subtle views on the rules for how potentials change.

The three classical force fields, g, E, and B, depend on the same quaternion potential, so there are only four degrees of freedom. With seven components to the three classical force fields, there must be three constraints between the fields. Two constraints are already familiar. The electric and magnetic field form a vector identity via Faraday's law. Assuming spacetime is simply connected, the no monopoles equation is another identity. A new constraint arises because both the force fields for gravity and electricity are even. It will be shown subsequently how the even force fields can partially constructively or destructively interfere with each other.

Unified Field Equations

In the standard approach to generating the Maxwell equations, a differential operator acts on the electromagnetic field strength tensor. A unified field hypothesis for an isolated source is proposed which involves a differential quaternion operator acting on the unified field:

$$4\pi\left(\rho,\bar{J}\right)^{*}=\left(\frac{\partial}{\partial t},\bar{\nabla}\right)^{*}\left(\frac{\partial}{\partial t},\bar{\nabla}\right)^{*}\left(\phi.\bar{A}\right)^{*}=$$

$$=\left(\frac{\partial^{2}\phi}{\partial t^{2}}-2\bar{\nabla}-\frac{\partial\bar{A}}{\partial t}-\bar{\nabla}-\bar{\nabla}\phi+\bar{\nabla}-\left(\bar{\nabla}\times\bar{A}\right),\right.$$

$$-2\bar{\nabla}\frac{\partial\phi}{\partial t}+\bar{\nabla}\left(\bar{\nabla}-\bar{A}\right)-\frac{\partial^{2}\bar{A}}{\partial t^{2}}-\bar{\nabla}\times\left(\bar{\nabla}\times\bar{A}\right)$$

$$\left.+\,2\bar{\nabla}\times\frac{\partial\bar{A}}{\partial t}+\bar{\nabla}\times\bar{\nabla}\phi\right).$$

This second order set of four partial differential equations has four unknowns so this is a complete set of field equations. Rewrite the equations above in terms of the classical force fields:

$$4\pi\left(\rho,\bar{J}\right)^{*}=\left(\frac{\partial g}{\partial t}+\bar{\nabla}.\bar{E}+\bar{\nabla}.\,\bar{B}-\bar{\nabla}g+\frac{\partial\bar{E}}{\partial t}-\bar{\nabla}\times\bar{B}+\frac{\partial\bar{B}}{\partial t}-\bar{\nabla}\times\bar{E}\right).$$

The unified field equations contain three of the four Maxwell equations explicitly:Gauss' law, the no magnetic monopoles law, and Ampere's law. Faraday's law is a vector identity, so it is still true implicitly. Therefore, a subset of the unified field equations contains a quaternion representation of the Maxwell equations. The justification for investigating the unified field equation hypothesis is due to the presence of the Maxwell equations.

There is a very simple relationship between Faraday's law

and the equation above. All that needs to be done is to subtract twice the time derivative of the magnetic field from both sides. What does this do to the 4-vector current density J? Now there is a current that transforms like a pseudo-current density, making the proposal more complete. The volume integral of this pseudo-current density is the total magnetic flux:

The unified field equation postulates a pseudo 3-vector current composed of the difference between the time derivative of the magnetic field and the curl of the electric field. The Aharonov-Bohm effect depends on the total magnetic flux to create changes seen in the energy spectrum. The volume integral of the time derivative of the magnetic field is a measure of the total magnetic flux.

The pseudo-current density is quite unusual, transforming differently under space inversion than the electric current density. One might imagine that a Lorentz transformation would shift this pseudo-current density into a pseudo-charge density. This does not happen however, because the vector identity involving the divergence of a curl still applies. The Aharonov Bohm phenomenon, first viewed as a purely quantum effect, may have a classical analogue in the unified field equations.

$$k\iiint \frac{\partial B}{\partial t} dv = \frac{e}{\hbar c}\Phi_B$$

The constants used here were chosen to suggest a connection to the Aharonov-Bohm. The Aharonov-Bohm effect depends on the total magnetic flux to create changes seen in the energy spectrum. A pseudo-current density is unusual, transforming differently under space inversion than the electric current density. One might imagine that a Lorentz transformation would shift this pseudo-current density into a pseudo-charge density.

This does not happen however, because the vector identity involving the divergence of a curl still applies. The Aharonov-Bohm phenomenon, first viewed as a purely quantum effect, may have a classical analogue in the unified field equations. The field equations involving the gravitational force field are dynamic and depend on four dimensions. This makes them likely to be consistent with special relativity. Since they are generated alongside the Maxwell equations, one can reasonably expect the differential equations will share many properties, with the ones involving the symmetric-matrix gravitational force field being more symmetric than those of the electromagnetic counterpart.

The unified source can be defined in terms of more familiar charge and current densities by separately setting the gravity or electromagnetic field equal to zero. In these cases, the source is due only to electricity or mass respectively. This leads to connections between the unified source, mass, and charge:

$$J = J_m \text{ iff } \vec{E} = \vec{B} = \vec{0}$$

$$J = J_e + \vec{J}_{AB} \text{ iff } g = 0.$$

It would be incorrect-but almost true-to say that the unified charge and current are simply the sum of the three: mass, electric charge, and the Aharanov-Bohm pseudo-current. These terms constructively interfere with each other, so they may not be viewed as being linearly independent.

Up to four linearly independent unified field equations can be formulated. A different set could be created by using the differential operator without taking its conjugate:

$$= 4\pi J = \Box^* \Box^* A^* =$$

$$= \left(\frac{\partial^2 \phi}{\partial t^2} + \vec{\nabla} - \vec{\nabla}\phi - \vec{\nabla} - \left(\vec{\nabla} \times \overline{A}\right), -\frac{\partial^2 \overline{A}}{\partial t^2} - \vec{\nabla}^2 \overline{A} - \vec{\nabla} \times \vec{\nabla}\phi \right)$$

$$= \left(\frac{\partial g}{\partial t} - \vec{\nabla}.\vec{E} - \vec{\nabla}.\vec{B}, \nabla g + \frac{\partial \vec{E}}{\partial t} + \vec{\nabla} \times \vec{B} + \frac{\partial \vec{B}}{\partial t} + \vec{\nabla} \times \vec{E} \right).$$

This is an elliptic equation. Since the goal of this work is a complete system of field equations, this may turn out to be an advantage. An elliptic equation combined with a hyperbolic one might more fully describe gravitational and electromagnetic waves from sources. Unlike the first set of field equations, the cross terms destructively interfere with each other. The elliptic field equation again contains three of four

Maxwell equations explicitly:Gauss' law, the no magnetic monopoles vector identity and Faraday's law. This time, Ampere's law looks different. To be consistent with Ampere's law, again a pseudo-current must be included. This may be the differential form of a classical Aharonov-Bohm effect. The only term that does not change between the two field equations is the one involving the dynamic gravitational force. This might be a clue for why this force is only attractive.

Solutions to the Unified Field Equations

All the solutions that have been worked out for the Maxwell equations will work with the unified field equations. For example, if the potential is static, the scalar equation for hyperbolic field equation is the Poisson equation. The unified equations are more informative, since any potential which is a solution to the scalar Poisson equation will also characterize the corresponding current.

The field equations of general relativity and the Maxwell equations both have vacuum solutions,such as plane wave solutions. The unified field equations do not have such a solution, other than a constant. Given historical tradition, this may seem like a deadly flaw. However, it may be something that is required for a final and complete theory. In a unified field theory, the gravitational part may be zero while the electrical part is not, and visa versa. Non-zero solutions are worth exploring. An inverse square potential plays an important role in both gravity and electromagnetism. Examine the scalar field involving the inverse interval squared:

$$\Box\Box^{*}\left(\frac{1}{t^2-x^2-y^2-z^2},\vec{0}\right)=\left(\frac{4\left(3t^2+x^2+y^2+z^2\right)}{\left(t^2-x^2-y^2-z^2\right)^3},\vec{0}\right)$$

This potential solves the Maxwell equations in the Lorentz gauge:

$$\Box\Box^{2}\left(\frac{1}{t^2-x^2-y^2-z^2},\vec{0}\right)=0$$

The non-zero part may have everything to do with

gravity. A plane wave solution does exist, but not for a pure vacuum. Instead, a plane wave solution exists with the constrain that the net current is zero for the elliptical field equations.

The field equations of general relativity and the Maxwell equations both have vacuum solutions. A vacuum solution for the unified field equation is apparent for the elliptical field equations:

$$\overline{A} = \left(\phi_0\ e^{\overline{k}\overline{k}-\omega t}, \overline{A}_0\ e^{\overline{k}\overline{k}-\omega t}\right)$$

The unified field equation will evaluate to zero if

$$\text{Scalar}\left(\left(\frac{\omega}{C}, \overline{K}\right)^2\right) = 0$$

The dispersion relation is an inverted distance, so it will depend on the metric. The same potential can also solve the hyperbolic field equations under different constraints and resulting dispersion equation. There were two reasons for not including the customary imaginary number "i" in the exponential of the potential.

First, it was not necessary, second, it would have created a complex-valued quaternion, and therefore is outside the domain of this paper. The important thing to realise is that vacuum solutions to the unified field equations exist whose dispersion equations depend on the metric.This is an indication that unifying gravity and electromagnetism is an appropriate goal.

CONSERVATION LAWS

Conservation of electric charge is implicit in the Maxwell equations. Is there also a conserved quantity for the gravitational field?Notice that the gravitational force field only appears in the quaternion scalar. The electromagnetic fields only appear in the 3-vector. This generates two types of constraints on the sources. No change in the electric source applies to the quaternion scalar. No change in the gravitational source applies to the 3-vector.

$$\text{Scalar}\left(\Box J_e^*\right)=\frac{\partial \rho e}{\partial t^2}+\overline{\nabla}-\overline{J}_e=0$$

$$\text{Scalar}\left(\Box \overline{J}_{AB}^*\right)=\nabla-\overline{J}_{AB}^*=0$$

$$\text{Vector}\left(\Box J_m^*\right)=-\frac{\partial \overline{J}_m}{\partial t}+\overline{\nabla}\rho_m-\overline{\nabla}\times\overline{J}_m=\overline{0}$$

The first equation is known as the continuity equation, and is the reason that electric charge is conserved. For a different inertial observer, this will appear as a conservation of electric current density. There is no source term for the Aharanov-Bohm current, and subsequently no conservation law. The 3-vector equation is a constraint on the mass current density, and is the reason mass current density is conserved. For a different inertial observer, the mass density is conserved.

TRANSFORMATIONS OF THE UNIFIED FORCE FIELD

The transformation properties of the unified field promise to be more intricate than either gravity or electromagnetism separately. What might be expected to happen under a Lorentz transformation? Gravity involves mass that is Lorentz invariant, so the field that generates it should be Lorentz invariant.

The electromagnetic field is Lorentz covariant. However, a transformation cannot do both perfectly. The reason is that a Lorentz transformation mixes a quaternion scalar with a 3-vector. If a transformation left the quaternion scalar invariant and the 3-vector covariant, the two would effectively not mix. The effect of unification must be subtle, since the transformation properties are well known experimentally.

Consider a boost along the x-axis. The gravitational force field is Lorentz invariant. All the terms required to make the electromagnetic field covariant under a Lorentz transformation are present, but covariance of the electromagnetic fields requires the following residual terms:

$$(\Box'^* A'^*)\text{Residual}=\left(0,\left(\gamma^2\beta-1\right)\frac{\partial A_x}{\partial t}+\left(\gamma^2-1\right)\frac{\partial \phi}{\partial x}\right.$$

$$-2\gamma\beta\left(\frac{\partial A_z}{\partial t}+\frac{\partial A_y}{\partial x}\right), 2\gamma\beta\left(\frac{\partial A_y}{\partial t}-\frac{\partial A_z}{\partial x}\right)\Bigg)$$

At this time, the correct interpretation of the residual term is unclear. These terms could be a velocity-dependent phase factor. If so, it might provide a test for the theory. The mechanics of the Lorentz transformation itself might require careful re-examination when so strictly confined to quaternion algebra. For a boost along the x-axis, if only the differential transformation is in the opposite direction, then the electromagnetic field is Lorentz covariant with the residual term residing with the gravitational field. The meaning of this observation is even less clear. Only relatively recently has DeLeo been able to represent the Lorentz group using real quaternions. The delay appears odd since the interval of special relativity is the scalar of the square of the difference between two events. In the real 4 × 4 matrix representation, the interval is a quarter of the trace of the square. Therefore, any matrix with a trace of one that does not distort the length of the scalar and 3-vector can multiply a quaternion without effecting the interval. One such class is 3-dimensional, spatial rotations. An operator that adds nothing to the trace but distorts the lengths of the scalar and 3-vector with the constraint that the difference in lengths is constant will also suffice. These are boosts in an inertial reference frame. Boosts plus rotations form the Lorentz group.

Three types of gauge transformations will be investigated: a scalar, a 3-vector, and a quaternion gauge field. Consider an arbitrary scalar field transformation of the potential:

$$\overrightarrow{\mathrm{A}} \rightarrow \overrightarrow{\mathrm{A}} = \overrightarrow{\mathrm{A}} - \square^{*}\lambda.$$

The electromagnetic fields are invariant under this transformation. An additional constraint on the gauge field is required to leave the gravitational force field invariant, namely that the scalar gauge field solves a homogeneous elliptical equation. From the perspective of this proposal, the freedom to choose a scalar gauge field for the Maxwell equations is due to the omission of the gravitational force field.

Transform the potential with an arbitrary 3-vector field:

$$\overline{A} \rightarrow \overline{A}' = \overline{A} - \square^{*} \overline{A}.$$

This time the gravitational force field is invariant under a 3-vector gauge field transformation. Additional constraints can be placed on the 3-vector gauge field to preserve a chosen electromagnetic invariant. For example, if the difference between the two electromagnetic fields is to remain invariant, then the 3-vector gauge field must be the solution to an elliptical equation. Other classes of invariants could be examined.

The scalar and 3-vector gauge fields could be combined to form a quaternion gauge field. This gauge transformation would have the same constraints as those above to leave the fields invariant. Is there any such gauge field? The quaternion gauge field can be represented the following way:

$$\overline{A} \rightarrow \overline{A}' = \overline{A} - \square^{*} \overline{\Lambda}.$$

If a force field is created by hitting this gauge transformation with a differential operator, then the gauge field becomes a unified field equation. Since vacuum solutions have been found for those equations, a quaternion gauge transformation can leave the field invariant.

FUTURE DIRECTIONS

The fields of gravity and electromagnetism were unified in a way consistent with Einstein's vision, not his technique. The guiding principles were simple but unusual:generate expressions familiar from electromagnetism using quaternions, striving to interpret any extra terms as being due to gravity. The first hypothesis about the unified field involved only a quaternion differential operator acting on a potential, no extra terms added by hand. It contained the typical potential representation of the electromagnetic field, along with a symmetric-matrix force field for gravity. The second hypothesis concerned a unified field equation formed by acting on the unified field with one more differential operator. All the Maxwell equations are included explicitly or implicitly. Additional terms suggested the inclusion of a classical representation of the Aharanov-Bohm effect. Four linearly independent unified field equations exist, but only the hyperbolic and elliptic cases were discussed. A large family of vacuum solutions exists, and will require future analysis to appreciate. To work within the guidelines of this paper, one should avoid solutions represented by complex-valued quaternions.

Why did this approach work? The hypothesis that initiated this line of research was that all events in spacetime could be represented by quaternions, no matter how the events were generated. This is a broad hypothesis, attempting to reach all areas covered by physics. Based on the equations presented in this paper, a logical structure can be constructed, starting from events. A set of events forms a pattern that can be described by a potential. The change in a potential creates a field. The change in field creates a field equation. The terms that do not change under differentiation of a field equation form conservation laws.

Chapter 5

Electrostatic Force

Electrostatics is associated with materials on which electrical charge moves only slowly (insulating materials) and with electrically isolated conductors. 'Insulation' and 'isolation' prevents easy migration of charge. So charges stay in place (are 'static'). It is the effects that the charges produce that are then important. The parameter of basic importance in electrostatics is 'charge'.

It is charge which gives rise to the electric fields generating forces that attract thin films and particles to surfaces and charge which gives rise to high voltages in low capacitance systems. High voltages can cause sparks. If these have sufficient energy they may ignite flammable gases, cause shocks to personnel, damage semiconductor devices and upset the operation of microelectronic equipment.

CHARGE

Static electricity arises as the separation of positive and negative charges at the interface between two dissimilar surfaces. If one or other of the surfaces prevent easy migration of charge, or the conductor on which they reside is isolated, then this charge is 'static' on the surface and remains available to influence the surroundings. 'Static' electricity can also arise on surfaces as trapped ions from the air. Static charges may be electrons, or positive, or negative ions–but they are in the basic units of electronic charge 1.602×10^{-19} coulomb. On a surface there are some 10^{19} atomic sites per m^{2}–so if the charge of even quite a small fraction of the surface atoms is changed then quite large quantities of charge are easily involved.

FORÇE BETWEEN CHARGES

The force between two charges, Q_1 and Q_2, is proportional to the quantities of charge, and inversely proportional to the square of the separation distance, *d*:

$$F = -Q_1\,Q_2/(4\pi\varepsilon_0 d^2)$$

This is Coulomb's Law. The force *F* is in Newtons, the quantities of charge Q_1 and Q_2 are in Coulombs and the separation distance *d* is in meters. The constant ε_0 is called the 'permittivity of free space' and has a value 8.854×10^{-12}. Charges of the same polarity repel each other and of opposite polarity attract.

Examples of electrostatic forces are the attraction between thin layers of plastic film and of fine dust particles to surfaces.

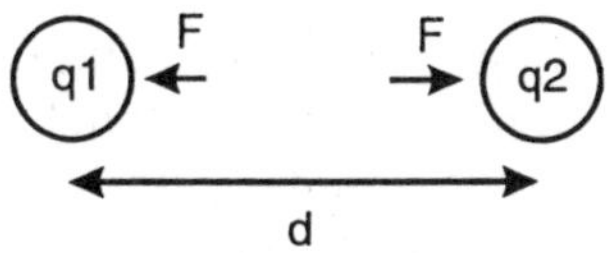

ELECTRIC FIELD

Around a charged body there is a force of attraction or repulsion for any other charges. This force, for a unit of charge, at any point is called the 'electric field', *E*, at that point. The direction of the field at any point depends on the direction of force on a positive charge there–and is hence a vector parameter. Diagrammatically it is useful to draw 'lines of force' to represent the electric field.

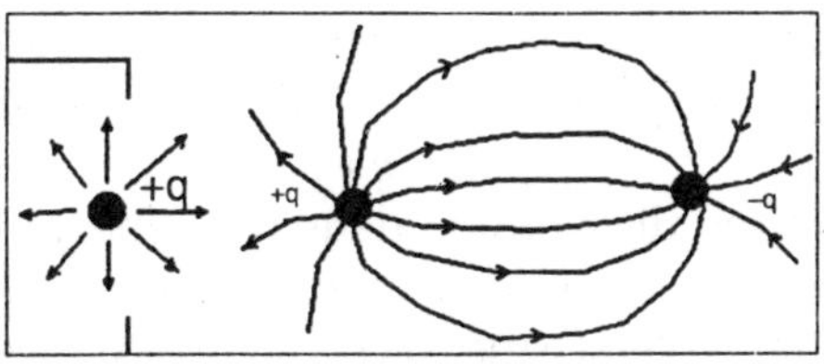

Fig. The Lines of Force the Stronger the Electric Field

From Coulombs Law the electric field strength *E* at a distance *d* from a charge *Q* will be:

$$E = Q/(4\pi\varepsilon_0 d^2)$$

The constant of proportionality $1/(4\pi\varepsilon_0)$ is, for historical reasons, defined in the MKS system of units by the relation:

$$1/(4\pi\varepsilon_0) = c^2 \times 10^{-7}$$

The constant ε_0 hence has a value 8.854×10^{-12} coulomb^2/(Newton m^2) or coulomb/(volt m). This is the 'permittivity of free space' noted above.

POTENTIAL

The 'potential' at a point is defined as the amount of work needed to bring a unit charge from infinity to that point. The potential difference between two points is then the work done to move a unit charge between these two points. The work done does not depend on the route followed so the potential is a scalar quantity.

The unit of potential is the 'volt'. The potential at a radial distance d from a single point charge Q is:

$$V = Q/(4\pi\varepsilon_0 d)$$

The electric field relates to potential as:

$$E = -\nabla V$$

ELECTRIC FLUX DENSITY

The electric field from an isolated point charge is radial. The field may conveniently be represented diagrammatically by lines of force. The 'flow' or flux of electric field across any spherical surface enclosing a point charge Q will then be Q/ε_o.

GAUSS' LAW

Gauss' Law states that the nett flux through any surface equals the nett charge enclosed within that surface.

$$\nabla.E = q/\varepsilon_0$$

For a uniform distribution density of charge ρ (C m^{-3}) in a sphere of radius r the total charge is $4/3\, \pi\, r^3\, \rho$. The outward flux will be $4\, \pi\, r^2\, E$. Hence the electric field at the sphere surface is:

$$E = \rho r\ /\ (3\varepsilon_0)$$

The maximum voltage at the centre of such a spherical distribution of charge is:

$$V = \rho r^2/(6\varepsilon_0)$$

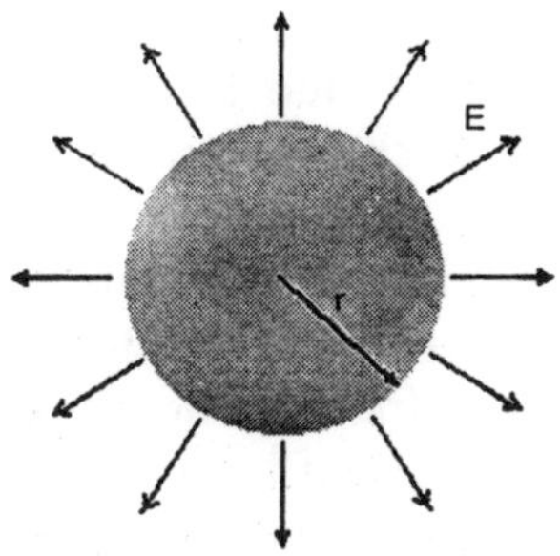

Similarly for a cylindrical distribution of charge density:

$$E = \rho r/(2\varepsilon_0) \text{ and } V = \rho r^2/(4\varepsilon_0)$$

INDUCED CHARGE

An electric field will have charge 'induced' on it with the density of charge proportional to the electric field. The charge induced on a conductor is of opposite polarity to that of the charge source. Opposite charges attract so there is an attraction between a charge and a conducting surface. For a plane conducting surface this 'image charge' is an equal distance behind the surface so calculation of the force involves a distance equal to twice the separation distance from the conducting surface. The electric field from each side of a sheet of uniform charge density σ (C m^{-2}) is:

$$E = \sigma/(2\varepsilon_0)$$

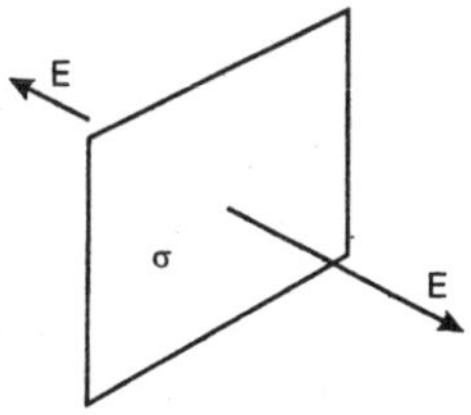

Fig. Electric Field from each side of a Sheet of Uniform Charge density

There can be no electric fields inside a conductor and the electric field outside is:

$$E = \sigma/\varepsilon_0$$

If a small conducting body (e.g. a small sphere) on an insulating handle is touched to a conducting body charge will be shared and the probe will acquire a charge dependent of the local electric field–related to the local surface charge density. If the amount of charge can be tested then we have a means to check if the conducting body was charged and to explore the density of charge over that body.

A 'gold leaf electroscope' indicates charge by repulsion between a fine gold leaf from it's nearby conducting support and gives an angle of the leaf indicating the level of charge. Sharing charge from a conducting probe with a gold leaf electroscope provides a good classical method to indicate and measure charge.

Such observations show that the electric field (and charge density) on a conducting body is inversely proportional to the local radius of curvature.

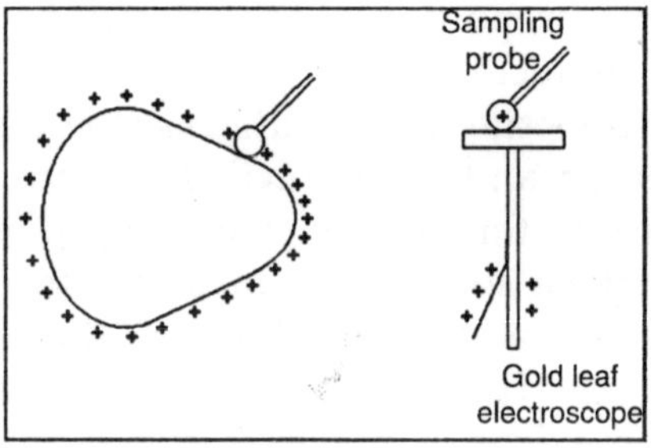

CAPACITANCE

For two conducting plates of area A separated by a distance d(m) with a voltage difference between them V the electric field between the plates is:

$$E = V/d$$

and the quantity of charge on each of the plates:

$$Q = A\,\sigma = \varepsilon_0\, AE = \varepsilon_0\, AV/d$$

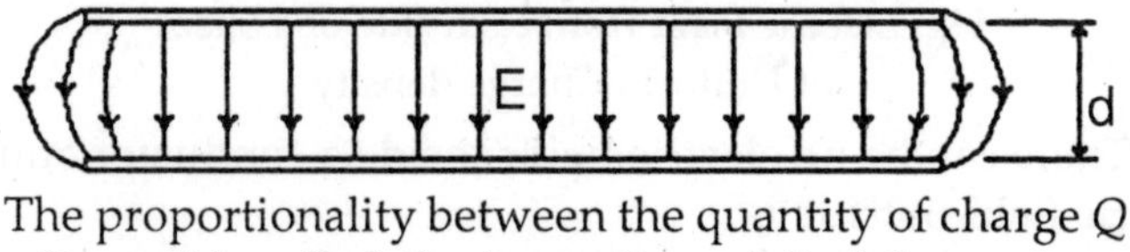

The proportionality between the quantity of charge Q and the voltage V is called the 'capacitance' C of the system. The unit of capacitance is the 'farad' with units of coulomb/volt. For the pair of parallel plates:

$$C = Q/V = \varepsilon_0 \, A/d$$

This relation is not quite accurate because the electric field is not uniform near the edges of the plates and extends outwards from them a bit.

For an isolated sphere of radius a (m) the capacitance is $4\pi\varepsilon_0 a$– for a radius of 10mm this is about 1 picofarad. The capacitance for a number of simple geometric arrangements (for instance concentric spheres, concentric cylinders, parallel rods and rods near plane conductors) can be caculated analytically. These can provide useful guidance for estimating capacitance in practical situations. Alternatively capacitance valus may be calculated by computer modelling calculations or, with suitable care, measured. The electrostatic energy U (joule) of a capacitor of C (farads) is:

$$U = 0.5\, CV^2$$

POISSON'S EQUATION

$$\nabla^2 V = -\rho / \varepsilon_0$$

where ñ is the volume density of charge (C m^{-3}). If no charges are present then Poisson's equation becomes Laplace's equation:

$$\nabla^2 V = 0$$

There are a number of geometric forms for which the above equations can be solved analytically. Two and three dimensional finite element and finite difference computer modelling programs are available to find potential and electric field distributions within practical geometric arrangements.

DIELECTRICS

If a block of conductor were placed between the capacitor plates the field in the remaining gap would be increased. A similar effect occurs with insulating materials, but because charges cannot move through or over the surface of insulators the material becomes 'polarised'. The atoms or molecules distort slightly under the electrical stress, polarise, and while these effects cancel out within the insulator they appear as surface charges on either side of the block of material. If an insulator is placed between the plates of a parallel plate capacitor the capacitance is increased.

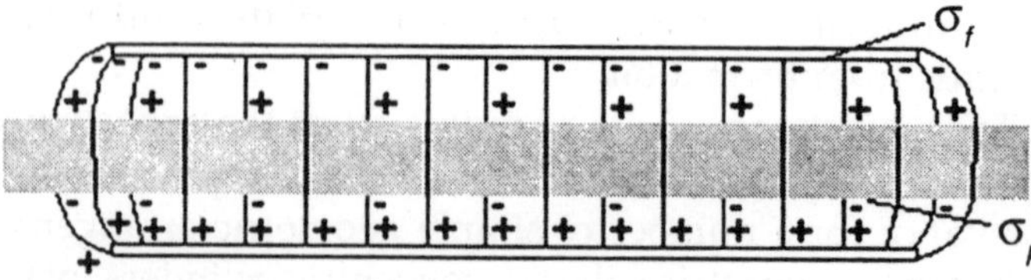

Applying Gauss's Law to the surface S the electric field in the insulator is:

$$E = (\sigma_f - \sigma_p)/\varepsilon_0$$

where σ_f and σ_p are the charge densities on the free surface of the capacitor plates and the surface of the insulator. The polarisation charges are proportional to the electric field and this is usually written:

$$P = X\,\varepsilon_0 E = \sigma_p$$

$$E = \sigma_f / \varepsilon_0(1 + X))$$

where X is the 'electric susceptibility' of the insulator/dielectric. This shows the reduction of the electric field within the insulator compared to that just outside.

The total charge on the capacitor plates is σ_f A so the capacitance by $Q = CV$ becomes for a capacitor fully filled by an insulator:

$$C = \sigma_f A/V = AE\ (0(1 + X))/(E\ d)$$

$$C = \varepsilon_0 A\ (1 + X)\ /\ d$$

The increase in capacitance for this fully filled capacitor is the factor (1 + X) and this is usually called the 'permittivity' of the material $-k$.

CHARGE MIGRATION AND DISSIPATION

Insulating materials have very few free electrical charges so only very small currents flow under the influence of an electric field. In more usual terms the volume 'resistivity' may be from 10^{12} to 10^{16} ohm m and the surface resistivity over 10^{10} ohms per square. Not only are there few charge carriers but their number and the velocity of their movement may depend in a non-linear way upon the electric field. In this situation 'resistivity' is not a very useful parameter for characterising materials. Of more direct practical importance in the control of static is the timescale for static charge dissipation. If the timescale for charge dissipation is less than the timescale for charge generation then no significant charge levels or surface voltages will arise and there will be no static problems.

For uniform materials, such as liquids, it is expected that the time for charge dissipation $\tau(\sigma)$ can be related to the volume resistivity ρ (ohms) and permittivity k as:

$$\tau = k\,\varepsilon_0\rho$$

CHARGE SEPARATION MECHANISMS

Contact between surfaces results in transfer of charge as electrons to match up energy levels of neighbouring atoms around the points of contact. The flow of charge continues until inhibited by the potential difference of the 'contact potential' of the surfaces. On separation the surfaces may retain the transferred charge.

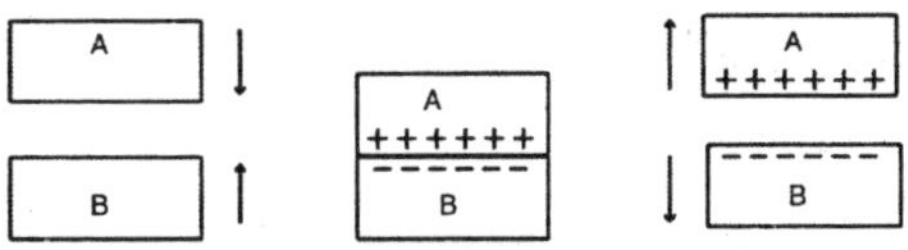

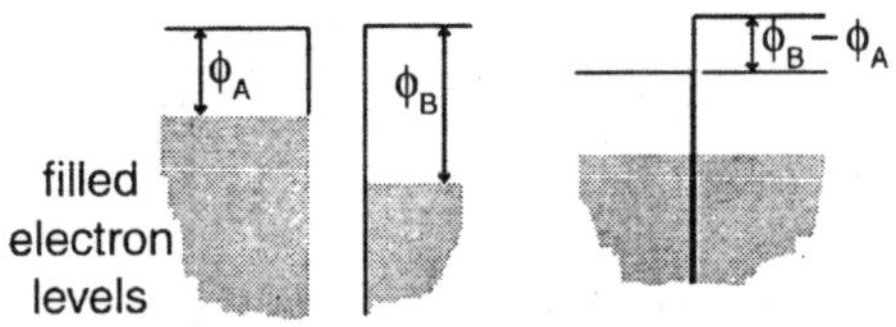

Charge transfer depends on differences in 'contact potential' and materials can be listed as a 'triboelectric series'. Charge separation depends on surface conditions. It is strongly affected by contamination and by the speed and pressure of rubbing actions.

Only in a few instances, for example in photocopiers and in some paint spray situation, are triboelectric effects used to control charging in a defined way. Charge separation also occurs when a liquid meets a conducting surface. Ions formed by dissociation in the fluid or from trace impurities will adsorb at the boundary surface. This creates the so-called 'double layer' at the boundary. An appreciable charge can be adsorbed to the surface–which is matched by charge in the fluid. With relatively insulating fluids the charge in the volume can be swept away by fluid flow as a 'streaming current' which can be in the range 10^{-10} to 10^{-7}A. A double layer also occurs when water breaks away from a surface–and this is responsible for the formation of the highly charged mists that can arise when high pressure water jets hit surfaces.

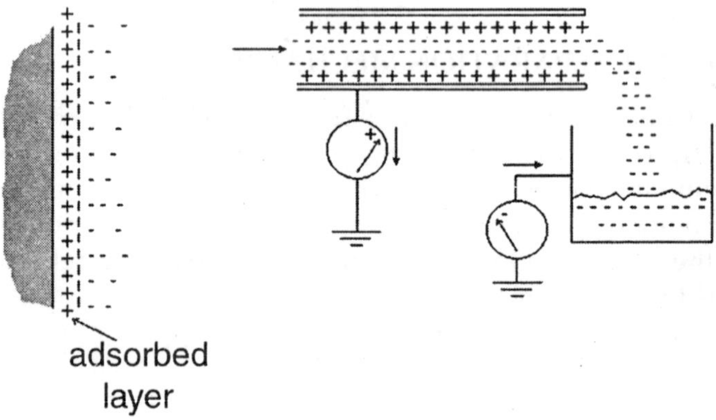

ELECTRICAL BREAKDOWN

If an electric field, in air at normal temperature and pressure, increases above about 3 MV m^{-1} then 'electrical breakdown' occurs. A loose electron or ion accelerates in the local electric field and if it picks up enough energy from the field before it hits another air molecule it will cause ionisation of that molecule. This can lead to further ionisation and rapid growth of the number of ions present. This multiplication of ionisation can destroy the insulating properties of the air and lead to passage of a high current accompanied by the emission of heat and light and sound–this is a spark discharge.

If the electric field exceeds the critical value for ion multiplication in just a localised region, for example near a sharp point, while the average electric field over the gap to nearby conductors is quite low, then a localised discharge takes place with a fairly low current (up to perhaps a mA) between the electrodes. This is a 'corona' discharge. If the electric field exceeds the breakdown strength of air over several mm, for example with an electrode of several mm radius of curvature, then the discharge involves more current and has a condensed luminous channel which does not reach to another electrode. This is a 'brush discharge'.

If the breakdown strength of air is exceeded near an electrode and also the average electric field over the gap to the other electrode exceeds about 500 kV m^{-1} then breakdown is likely to occur as a spark discharge between the electrodes. The voltage for electrical breakdown depends on the geometry of the gap, on the gas and on the pressure. As the pressure or gap is reduced the breakdown voltage decreases to a minimum and then rises-following the Paschen Law curve. For air the minimum is about 300 volts.

If a high electric field arises at the surface of a fluid the surface will distort and break up. This occurs at a lower electric field than that for air breakdown.

COULOMB'S LAW OF ELECTROSTATIC FORCES

CHARGE INTERACTIONS BY COULOMB'S LAW

Electrical interactions between charges govern much of physics, chemistry, and biology. They are the basis for chemical bonding, weak and strong. Salts dissolve in water to form the solutions of charged ions that cover three-quarters of the Earth's surface. Salt water forms the working fluid of living cells. pH and salts regulate the associations of proteins, DNA, cells, and colloids, and the conformations of biopolymers.

Nervous systems would not function without ion fluxes. Electrostatic interactions are also important in batteries, corrosion, and electroplating. Charge interactions obey Coulomb's law. When more than two charged particles interact, the energies are sums of coulombic interactions. With these tools you can determine the electrostatic force exerted by one charged object on another, as when an ion interacts with a protein, DNA molecule, or membrane, or when a charged polymer changes conformation. Coulomb's law was discovered in careful experiments by H Cavendish (1731–1810), J Priestley (1733–1804), and CA Coulomb (1736–1806) on macroscopic objects such as magnets, glass rods, charged spheres, and silk cloths. But Coulomb's law applies to a wide range of size scales, including atoms, molecules and biological cells. It states that the interaction energy *u(r)* between two charges in a vacuum is

$$u(r) = e\frac{q_1 q_2}{r},$$

where q_1 and q_2 are the magnitudes of the two charges, r is the distance separating them, and $e = 1/4\pi\varepsilon_0$ is a proportionality constant.

CHARGE INTERACTIONS BY LONG-RANGED

Uncharged atoms interact with each other through *short-ranged* interactions. That is, attractions between uncharged atoms are felt only when the two atoms are so close to each other that they nearly contact. Short-ranged interactions typically diminish with distance as r^{-6} it suffices to note that charge interactions are very different: they are long-ranged. Coulombic interactions diminish with distance as r^{-1}. Counting lattice contacts is no longer sufficient.

The difference between short-ranged and long-ranged interactions is much more profound than it may seem. The mathematics and physics of longranged and short-ranged interactions are very different.

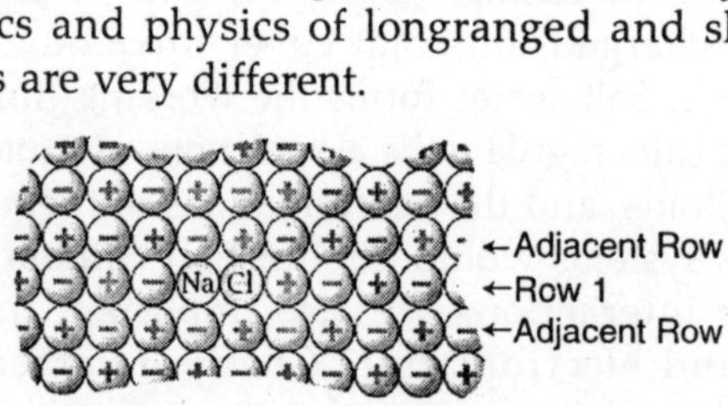

Fig. Crystalline Sodium Chloride Packs in a Cubic Array of Alternating Sodium and Chloride Ions

A particle that interacts through short-ranged interactions feels only its nearest neighbors, so system energies can be computed by counting the nearest-neighbour contacts. But when interactions involve charges, more distant neighbors contribute to the energies, so our methods of summing energies must become more sophisticated.

Electrostatic interactions are strong as well as long-ranged. The electrostatic repulsion is about 1036 times stronger than the gravitational attraction between two protons. Strong, short-ranged nuclear forces are required to hold the protons together against their electrostatic repulsions in atomic nuclei. When charge builds up on a macroscopic scale, violent events may result, such as lightning and thunder storms, and explosions in oil tankers. Because charge interactions are so strong, matter does not sustain charge separations on macroscopic scales: macroscopic matter is neutral.

The transfer of less than a thousand charges is enough to cause the static electricity that you observe when you rub your shoes on a carpet. In solution, too, imbalances are usually less than thousands of charges, but because this is one part in 1021, it means that you can assume that solutions have charge neutrality—no net charge of one sign—to a very high degree of approximation. Although macroscopic charge separations aren't common in ordinary solutions, charge separations on microscopic scales.

CHARGE INTERACTIONS BY WEAKER IN MEDIA

Charges interact more weakly when they are in liquids than when they are in a vacuum. To varying degrees, liquids can be *polarized*. This means that if two fixed charges, say a negative charge A and a positive charge B, are separated by a distance r in a liquid, then the intervening molecules of the liquid tend to reorient their charges.

The positive charges in the medium orient toward A, and the negative charges toward B, to shield and weaken the interaction between A and B. This weakening of the coulombic interactions between A and B is described by the *dielectric constant* of the medium. Media that can be readily polarized shield charge interactions strongly: they have high dielectric constants.

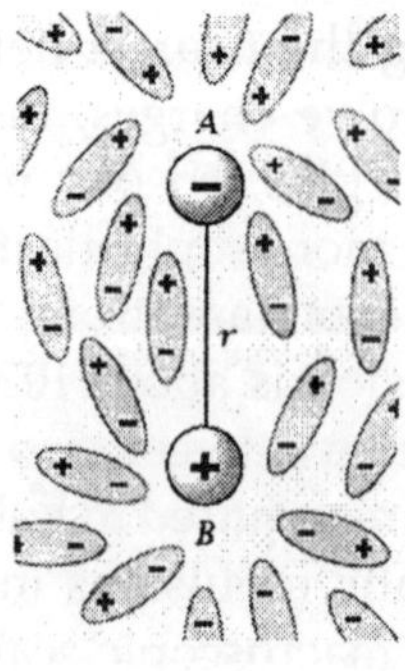

Fig. Fixed Charges *A* and *B* cause Dipoles in the Surrounding Liquid to Orient partly, which causes some *Shielding*, a Reduction of the Interaction between *A* and *B*

The medium as a *polarizable isotropic continuum*. In treating a medium as a continuum, we neglect its atomic structure and focus on its larger-scale properties. In treating a medium as *isotropic*, we assume that its polarizability is the same in all directions. By treating a medium as *polarizable*, we assume that the charge redistributes in response to an electric field, even if the medium is neutral overall. The more polarizable the medium, the greater is the reduction of an electrostatic field across it. The reduction factor is the *dielectric constant D*, a scalar dimensionless quantity. When charges q_1 and q_2 are separated by a distance r in a medium of dielectric constant D, their coulombic interaction energy is

$$u(r) = \frac{eq_1q_2}{Dr}.$$

Coulombic interactions are traditionally described as an energy *u(r)*, but when dielectric constants are involved, *u(r)* is a free energy, because the dielectric constant is temperature dependent. The polarizability of a medium arises from several factors. First, if the medium is composed of molecules that have a *permanent dipole moment*, a positive charge at one end and a negative charge at the other, then applying an electric field tends to orient the dipoles of the medium in a direction that opposes the field.

Second, the polarizability of a medium can arise from the

polarizabilities of the atoms or molecules composing it, even if they have no permanent dipole moment. Atoms or molecules that lack permanent dipoles have electronic *polarizability*, a tendency of nuclear or electronic charge distributions to shift slightly within the atom, in response to an electric field. The electronic polarizabilities of hydrocarbons and other nonpolar substances are the main contributors to their dielectric constants ($D \approx 2$).

Third, polarizabilities can also arise from networks of hydrogen bonds in liquids such as water. If there are many alternative hydrogen-bond donors and acceptors, a shift in the pattern of hydrogen bonding is not energetically costly, so the networks of hydrogen bonds in hydrogen-bonded liquids are labile when electric fields are applied.

THE BJERRUM LENGTH

To show the effect of the polarizability of the medium, let's first compute a useful quantity called the *Bjerrum length –B*, defined as the charge separation at which the coulomb energy between ions *u(r)* just equals the thermal energy *RT*.

Substitute ℓ_B for *r* and *RT* for *u* in $u(r) = eq_1q_2/Dr$ and solve for ℓ_B for single charges $q_1 = q_2 = e$:

$$\ell_B = \frac{ee^2N}{DRT}.$$

For a temperature $T = 298.15$, in a vacuum or air ($D = 1$),

$$\ell_B = \frac{1.386 \times 10^{-4}\,\text{J m mol}^{-1}}{(8.314\,\text{J K}^{-1}\text{mol}^{-1} \times 298.15K)} = 560\,\mathring{A}.$$

560 Å is a relatively long distance, many times the diameter of an atom. When two charges are much closer together than this distance, they feel a strong repulsive or attractive interaction, much larger than the thermal energy *RT*. But when two charges are much further apart than this distance, their interactions are weaker than *RT*, and they are more strongly directed by Brownian motion than by their coulombic interactions.

Dividing $u(r) = eq_1q_2/Dr$ by $(RT = eq_1q_2/D\ell_B)$ gives

$$\frac{u}{RT} = \frac{\ell_B}{r}.$$

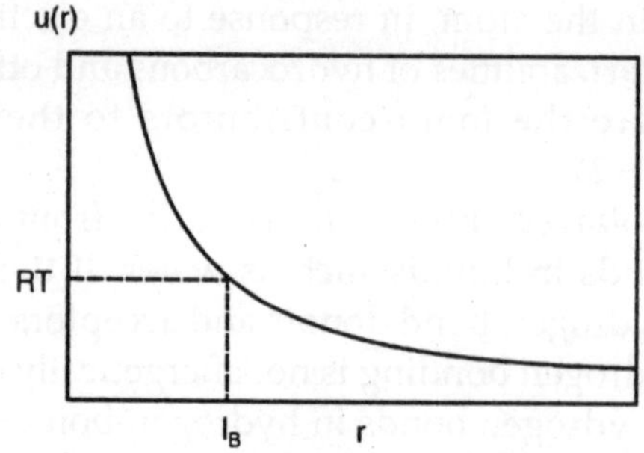

Fig. The Bjerrum length *B* is defined as the distance *r* at which two unit charges interact with an energy *u(r)* equal to *RT*

So far, our calculations for ionic interactions have involved simple arrangements of charges. When several charges are involved, or when charges have complex spatial arrangements, more sophisticated methods are needed to treat them. We now develop some useful tools that lead to Gauss's law and Poisson's equation.

ELECTROSTATIC FORCES ADD LIKE VECTORS

Vectors and Coulomb's Law

Coulomb's law may be expressed either in terms of energy $u(r) = eq_1q_2/Dr$, or force $f = -(\partial u/\partial r) = eq_1q_2/Dr^2$.

The difference is that energies are scalar quantities, which add simply, while forces are vectors that add in component form. Either way, an important fundamental law is the superposition principle: electrostatic energies and forces are both additive. The total electrostatic force on a particle is the vector sum of the electrostatic forces from all other particles.

$$f = \frac{eq_Aq_B}{Dr^2} \cdot \frac{r}{r},$$

Where *r*/*r* indicates a vector pointing in the direction of vector *r*, along the line connecting the particles, but with unit length. *f* is the force acting in the same direction.

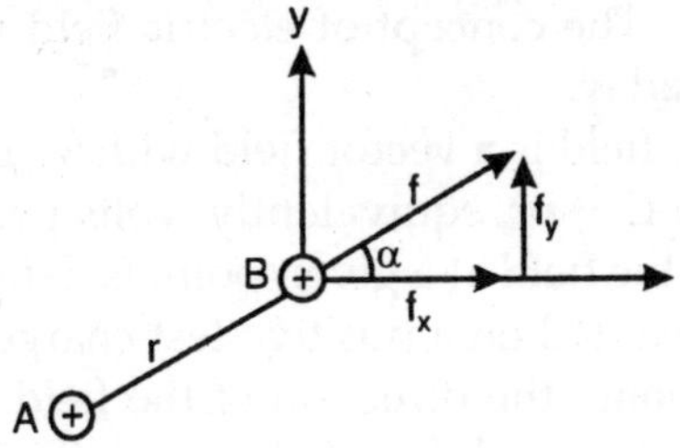

Fig. The Electrostatic Force *f* of fixed charge *A* on test charge *B* is a Vector with Components *fx* and *fy* along the *x*- and *y*-axes, respectively

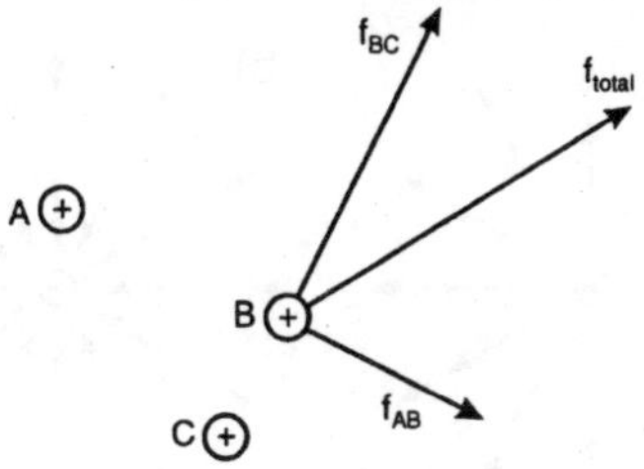

Fig. The Force on the Test Charge *B* is the Vector Sum from the two Fixed Charges *A* and *C*

The component of the force *f* in the *x*-direction is described by using the component of the vector *r* between the charges, *rx* = *r* cos α, where α is the angle between *f* and the *x*-axis. The *x*-direction component of the force is

$$f_x = \left(\frac{q_A q_B}{4\pi\varepsilon_0 D r^2}\right)\cos\alpha.$$

The *y*-direction component is

$$f_y = \left(\frac{q_A q_B}{4\pi\varepsilon_0 D r^2}\right)\sin\alpha.$$

ELECTRIC FIELD

In physics, the space surrounding an electric charge or in the presence of a time-varying magnetic field has a property called an electric field (that can also be equated to electric flux density). This electric field exerts a force on other electrically charged objects. The concept of electric field was introduced by Michael Faraday.

The electric field is a vector field with SI units of newtons per coulomb ($N\ C^{-1}$) or, equivalently, volts per meter ($V\ m^{-1}$). The strength of the field at a given point is defined as the force that would be exerted on a positive test charge of +1 coulomb placed at that point; the direction of the field is given by the direction of that force. Electric fields contain electrical energy with energy density proportional to the square of the field intensity. The electric field is to charge as gravitational acceleration is to mass and force density is to volume.

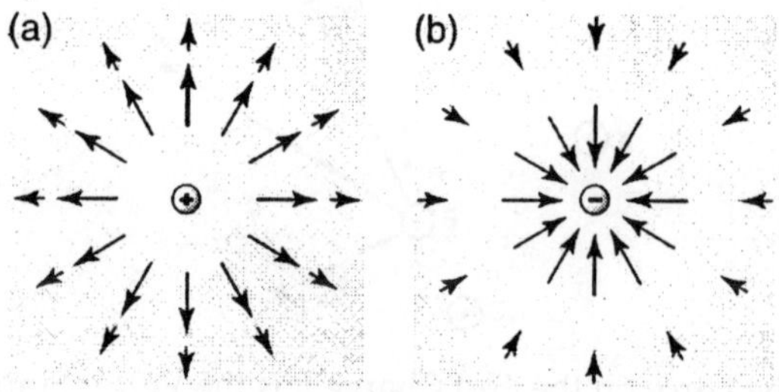

Fig. The Electric Field is indicated by Vectors Pointing (a) away from a Positive Charge and (b) Toward a Negative Charge. The Field Represents the Force Per Unit Charge that acts on a Positive Test Charge Put at Each Particular Position in Space

A moving charge has not just an electric field but also a magnetic field, and in general the electric and magnetic fields are not completely separate phenomena; what one observer perceives as an electric field, another observer in a different frame of reference perceives as a mixture of electric and magnetic fields.

For this reason, one speaks of "electromagnetism" or "electromagnetic fields." In quantum mechanics, disturbances in the electromagnetic fields are called photons, and the energy of photons is quantized. A stationary charged particle in an electric field experiences a force proportional to its charge given by the equation,

$$F = q\left[-\nabla\phi - \frac{\partial A}{\partial t}\right]$$

where the magnetic flux density is given by,

$$B = \nabla \times A$$

and where $-\nabla\phi$ is the Coulomb force.

Electric charge is a characteristic of some subatomic particles, and is quantized when expressed as a multiple of the so-called elementary charge e. Electrons by convention have a charge of –1, while protons have the opposite charge of +1. Quarks have a fractional charge of –1/3 or +2/3. The antiparticle equivalents of these have the opposite charge. There are other charged particles.

In general, same-sign charged particles repel one another, while different-sign charged particles attract. This is expressed quantitatively in Coulomb's law, which states the magnitude of the repelling force is proportional to the product of the two charges, and weakens proportionately to the square of the distance.

The electric charge of a macroscopic object is the sum of the electric charges of its constituent particles. Often, the net electric charge is zero, since naturally the number of electrons in every atom is equal to the number of the protons, so their charges cancel out. Situations in which the net charge is non-zero are often referred to as static electricity.

Even when the net charge is zero, it can be distributed non-uniformly (e.g., due to an external electric field), and then the material is said to be polarized, and the charge related to the polarization is known as bound charge (while the excess charge brought from outside is called free charge).

An ordered motion of charged particles in a particular direction (in metals, these are the electrons) is known as electric current. The discrete nature of electric charge was proposed by Michael Faraday in his electrolysis experiments, then directly demonstrated by Robert Millikan in his oil-drop experiment.

The SI unit for quantity of electricity or electric charge is the coulomb, which represents approximately 1.60×10^{19}

elementary charges (the charge on a single electron or proton). The coulomb is defined as the quantity of charge that has passed through the cross-section of an electrical conductor carrying one ampere within one second. The symbol Q is often used to denote a quantity of electricity or charge.

The quantity of electric charge can be directly measured with an electrometer, or indirectly measured with a ballistic galvanometer. Formally, a measure of charge should be a multiple of the elementary charge e (charge is quantized), but since it is an average, macroscopic quantity, many orders of magnitude larger than a single elementary charge, it can effectively take on any real value. In some contexts it is meaningful to speak of fractions of a charge; e.g. in the charging of a capacitor. If the charged particle can be considered a point charge, the electric field is defined as the force it experiences per unit charge:

$$E = \frac{F}{q}$$

where

F is the electric force experienced by the particle

q is its charge

E is the electric field wherein the particle is located

Taken literally, this equation only defines the electric field at the places where there are stationary charges present to experience it. The force exerted by another charge q will alter the source distribution, which means the electric field in the presence of q differs from itself in the absence of q. However, the electric field of a given source distribution remains defined in the absence of any charges with which to interact.

This is achieved by measuring the force exerted on successively smaller *test charges* placed in the vicinity of the source distribution. By this process, the electric field created by a given source distribution is defined as the limit as the test charge approaches zero of the force per unit charge exerted there upon.

$$E = \lim_{q \to 0} \frac{F}{q}$$

This allows the electric field to be dependent on the source distribution alone.

As is clear from the definition, the direction of the electric field is the same as the direction of the force it would exert on a positively-charged particle, and opposite the direction of the force on a negatively-charged particle. Since like charges repel and opposites attract the electric field tends to point away from positive charges and towards negative charges

Electric Fields Have Fluxes

The electric field flux has the important general property that it is independent of the size and shape of any balloon that fully surrounds all the charges, as we will show. This generality leads to some important methods for simplifying the problem of computing complex electrostatic fields. Positive and negative charges are the sources and sinks of electrical fields.

Consider some constellation of positive and negative charges fixed at certain positions in space. Just as we did for fluids we invent an imaginary balloon that we place around the charges. We want to determine the flux of the electric field through the imaginary balloon surface. The electric field flux Φ is defined as the integral of D times the electric field E over all the surface area elements ds of the balloon, where D is the dielectric constant. Because E is proportional to $1/D$ in a uniform medium the flux Φ is independent of D.

$$\Phi = \int_{\text{surface}} DE.ds,$$

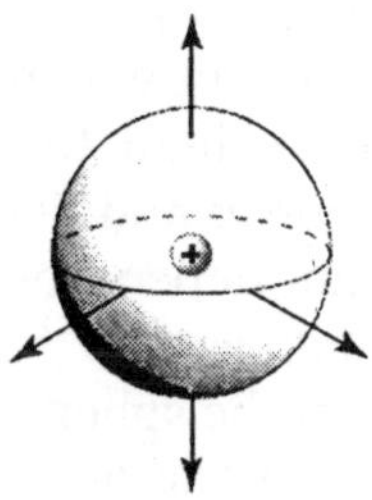

Fig. The Vector Electric Field E is Normal to a Sphere Surrounding a Point Charge located at the Centre of the Sphere

Equation is general and applies to any constellation of fixed charges, and to any shape and position of an imaginary balloon around them. Many problems of interest have a special simplicity or symmetry that allows you to simplify Equation. For example, suppose that the charge constellation is just a single point charge q and that the imaginary balloon is just a sphere of radius r centered on the point charge. We are interested in computing the electric field flux from the point charge through the imaginary spherical surface.

The electrostatic field of a point charge is spherically symmetrical—the magnitude of the field E is the same through the north pole as it is through the south pole or through the equator, or through any other patch of surface of a given area. So you can drop the vector notation and express the field as $E(r)$, which depends only on r. Because E doesn't differ from one surface patch to another, the integral in Equation becomes a simple product.

That is, E is parallel to ds everywhere, so $E \cdot ds = Eds$ and the flux integral becomes $\Phi = DE \int ds = DE(r) \times$ (total area of sphere). Use $E(r) = q/(4\pi\varepsilon_0 Dr^2)$ from Equation and $4\pi r^2$ for the surface area of a sphere, to get

$$\Phi = DE(r)(4\pi r^2)$$

$$= \left(\frac{q}{4\pi\varepsilon_0 r^2}\right)(4\pi r^2)$$

$$= \frac{q}{\varepsilon_0},$$

where q is the charge at the centre of the sphere. Two profound implications of Equation lead to a general method for calculating electrostatic forces and fields. First, the right-hand side of Equation does not depend on r. For the electric flux through a spherical container, the r^{-2} dependence of E cancels with the r^2 dependence of the spherical surface area. Because of this cancelation, the electric field flux out of a sphere is independent of its radius. It doesn't matter how large you choose your imaginary balloon to be, the flux is the same through spherical balloons of any radius.

POISSON'S EQUATION

In mathematics, Poisson's equation is a partial differential equation with broad utility in electrostatics, mechanical engineering and theoretical physics. It is named after the French mathematician, geometer and physicist Siméon-Denis Poisson. The Poisson equation is

$$\Delta\varphi = f$$

where Δ is the Laplace operator, and f and ö are real or complex-valued functions on a manifold. When the manifold is Euclidean space, the Laplace operator is often denoted as ∇^2 and so Poisson's equation is frequently written as

$$\nabla^2\varphi = f$$

In three-dimensional Cartesian coordinates, it takes the form

$$\left(\frac{\partial^2}{\partial x^2}+\frac{\partial^2}{\partial y^2}+\frac{\partial^2}{\partial z^2}\right)\varphi(x,y,z) = f(x,y,z).$$

For vanishing f, this equation becomes Laplace's equation

$$\Delta\varphi = 0$$

The Poisson equation may be solved using a Green's function; a general exposition of the Green's function for the Poisson equation is given in the article on the screened Poisson equation. There are various methods for numerical solution. The relaxation method, an iterative algorithm, is one example.

ELECTROSTATICS

One of the principal cornerstones of electrostatics is the posing and solving of problems that are described by the Poisson equation. Finding φ for some given f is an important practical problem, since this is the usual way to find the electric potential for a given charge distribution.

The derivation of Poisson's equation in electrostatics follows. SI units are used and Euclidean space is assumed.

Starting with Gauss' law for electricity in a differential control volume, we have:

$$\nabla . D = \rho$$

∇ means to take the divergence.

D is the electric displacement field.

ρ is the charge density.

Assuming the medium is linear, isotropic, and homogeneous, then:

$$D = \varepsilon E = \varepsilon_r \varepsilon_0 E$$

ε is the permittivity of the medium.

E is the electric field.

ε_0 is the vacuum permittivity.

ε_r is the relative permittivity of the medium.

By substitution and division, we have:

$$\nabla . E = \frac{\rho}{\varepsilon_r \varepsilon_0}$$

In the absence of a changing magnetic field, B, Faraday's law of induction gives:

$$\nabla \times E = -\frac{\partial B}{\partial t} = 0$$

$\times$ is the cross product.

t is time.

Since the curl of the electric field is zero, it is defined by a scalar electric potential field, φ

$$E = -\nabla \varphi$$

Eliminating E by substitution, we have a form of the Poisson equation:

$$\nabla . \nabla \varphi = \nabla^2 \varphi = -\frac{\rho}{\varepsilon_r \varepsilon_0}$$

Solving Poisson's equation for the potential requires knowing the charge density distribution. If the charge density is zero, then Laplace's equation results. If the charge density follows a Boltzmann distribution, then the Poisson-Boltzmann equation results.

The Poisson-Boltzmann equation plays a role in the development of the Debye-Hückel theory of dilute electrolyte solutions. The magnetic field not varying in time, the same

Poisson equation arises even if it does vary in time, as long as the Coulomb gauge is used. However, in this more general context, computing φ is no longer sufficient to calculate E, since the latter also depends on the magnetic vector potential, which must be independently computed.

POTENTIAL OF A GAUSSIAN CHARGE DENSITY

If there is a spherically symmetric Gaussian charge density $\rho(r)$:

$$\rho(r) = \frac{Q}{\sigma^3\sqrt{2\pi}^3} e^{-r^2/(2\sigma^2)},$$

where Q is the total charge, then the solution $\varphi(r)$ of Poisson's equation with $\varepsilon_r = 1$,

$$\nabla^2\varphi = -\frac{\rho}{\varepsilon_0},$$

is given by

$$\varphi(r) = \frac{1}{4\pi\varepsilon_0}\frac{Q}{r} erf\left(\frac{r}{\sqrt{2}\sigma}\right).$$

where erf(x) is the error function. This solution can be checked explicitly by a careful manual evaluation of $\nabla^2\varphi$. Note that, for r much greater than σ, erf(x) approaches unity and the potential $\phi(r)$ approaches the point charge potential $\frac{1}{4\pi\varepsilon_0}\frac{Q}{r}$, as one would expect. The erf function approaches 1 extremely fast as its argument increase; in practice for $r > 3\sigma$ the relative error is smaller than 1/1000.

LAPLACE'S EQUATION

Laplace's equation is a partial differential equation named after Pierre-Simon Laplace who first studied its properties. The solutions of Laplace's equation are important in many fields of science, notably the fields of electromagnetism, astronomy, and fluid dynamics, because they describe the behaviour of electric, gravitational, and fluid potentials. The general theory of solutions to Laplace's equation is known as potential theory. In the study of heat conduction, the Laplace equation is the steady-state heat equation. In three dimensions, the problem is to find twice-differentiable real-valued functions, φ of real variables, x, y, and z, such that

$$\frac{\partial^2 \varphi}{\partial x^2}+\frac{\partial^2 \varphi}{\partial y^2}+\frac{\partial^2 \varphi}{\partial z^2}=0.$$

This is often written as

$$\nabla^2 \varphi = 0$$

or

$$\text{div grad } \varphi = 0,$$

where div is the divergence, and grad is the gradient, or

$$\Delta\varphi = 0,$$

where Δ is the Laplace operator.

Solutions of Laplace's equation are called harmonic functions.

If the right-hand side is specified as a given function, $f(x, y, z)$, i.e.

$$\Delta\varphi = f$$

then the equation is called "Poisson's equation." Laplace's equation and Poisson's equation are the simplest examples of elliptic partial differential equations. The partial differential operator, ∇^2, or Δ, (which may be defined in any number of dimensions) is called the Laplace operator, or just the Laplacian.

BOUNDARY CONDITIONS

The Dirichlet problem for Laplace's equation consists of finding a solution φ on some domain D such that φ on the boundary of D is equal to some given function. Since the Laplace operator appears in the heat equation, one physical interpretation of this problem is as follows: fix the temperature on the boundary of the domain and wait until the temperature in the interior doesn't change anymore; the temperature distribution in the interior will then be given by the solution to the corresponding Dirichlet problem.

The Neumann boundary conditions for Laplace's equation specify not the function φ itself on the boundary of *D*, but its normal derivative. Physically, this corresponds to the construction of a potential for a vector field whose effect is known at the boundary of *D* alone.

Solutions of Laplace's equation are called harmonic functions; they are all analytic within the domain where the equation is satisfied. If any two functions are solutions to Laplace's equation (or any linear homogenous differential equation), their sum (or any linear combination) is also a solution. This property, called the principle of superposition, is very useful, e.g., solutions to complex problems can be constructed by summing simple solutions.

LAPLACE EQUATION IN TWO DIMENSIONS

The Laplace equation in two independent variables has the form

$$\varphi_{xx} + \varphi_{yy} = 0.$$

Analytic Functions

The real and imaginary parts of a complex analytic function both satisfy the Laplace equation. That is, if $z = x + iy$, and if

$$f(z) = u(x, y) + iv(x, y),$$

then the necessary condition that $f(z)$ be analytic is that the Cauchy-Riemann equations be satisfied:

$$u_x = v_y, \; v_x = -u_y.$$

It follows that

$$u_{yy} = (-v_x)_y = -(v_y)_x = -(u_x)_x.$$

Therefore u satisfies the Laplace equation. A similar calculation shows that v also satisfies the Laplace equation.

Conversely, given a harmonic function, it is the real part of an analytic function, $f(z)$ (at least locally). If a trial form is

$$f(x) = \varphi(x, y) + i\psi(x, y),$$

then the Cauchy-Riemann equations will be satisfied if we set

$$\psi_x = \psi_y, \; \psi_y = \varphi x.$$

This relation does not determine ø, but only its increments:

$$d\psi = -\varphi_y dx + \varphi_x\, dy.$$

The Laplace equation for φ implies that the integrability condition for ψ is satisfied:

$$\psi_{xy} = \psi_{yx}.$$

and thus ψ may be defined by a line integral. The integrability condition and Stokes' theorem implies that the value of the line integral connecting two points is independent of the path. The resulting pair of solutions of the Laplace equation are called conjugate harmonic functions. This construction is only valid locally, or provided that the path does not loop around a singularity. For example, if r and θ are polar coordinates and

$$\varphi = \log r,$$

then a corresponding analytic function is

$$f(x) = \log z = \log r + i\theta.$$

However, the angle θ is single-valued only in a region that does not enclose the origin.

The close connection between the Laplace equation and analytic functions implies that any solution of the Laplace equation has derivatives of all orders, and can be expanded in a power series, at least inside a circle that does not enclose a singularity. This is in sharp contrast to solutions of the wave equation, which generally have less regularity.

There is an intimate connection between power series and Fourier series. If we expand a function f in a power series inside a circle of radius R, this means that

$$f(z) = \sum_{n=0}^{\infty} c_n z^n,$$

with suitably defined coefficients whose real and imaginary parts are given by

$$c_n = a_n + ib_n.$$

Therefore

$$f(z) = \sum_{n=0}^{\infty} [a_n r^n \cos n\theta - b_n r^n \sin n\theta]$$

$$+ \, i\sum_{n=1}^{\infty}[a_n r^n \sin n\theta + b_n r^n \cos n\theta],$$

which is a Fourier series for *f*.

Fluid Flow

Let the quantities *u* and *v* be the horizontal and vertical components of the velocity field of a steady incompressible, irrotational flow in two dimensions. The condition that the flow be incompressible is that

$$u_x + v_y = 0,$$

and the condition that the flow be irrotational is that

$$\nabla \times V = v_x - u_y = 0.$$

If we define the differential of a function ψ by

$$d\psi = v\,dx - u\,dy,$$

then the incompressibility condition is the integrability condition for this differential: the resulting function is called the stream function because it is constant along flow lines. The first derivatives of ψ are given by

$$\psi x = v, \quad \psi y = -u,$$

and the irrotationality condition implies that ψ satisfies the Laplace equation. The harmonic function ψ that is conjugate to ψ is called the velocity potential. The Cauchy-Riemann equations imply that;

$$\varphi_x = -u, \; \varphi_y = -v.$$

Thus every analytic function corresponds to a steady incompressible, irrotational fluid flow in the plane. The real part is the velocity potential, and the imaginary part is the stream function.

According to Maxwell's equations, an electric field (*u*,*v*) in two space dimensions that is independent of time satisfies

$$\nabla \times (u, v) = v_x - u_y = 0,$$

and

$$\nabla.(u,v)=\rho$$

where ρ is the charge density. The first Maxwell equation is the integrability condition for the differential

$$d\varphi = -u\,dx - v\,dy,$$

so the electric potential φ may be constructed to satisfy

$$\varphi_x = -u,\ \varphi_y = -v.$$

The second of Maxwell's equations then implies that

$$\varphi_{xx} + \varphi_{yy} = -\rho,$$

which is the Poisson equation.

It is important to note that the Laplace equation can be used in three-dimensional problems in electrostatics and fluid flow just as in two dimensions.

LAPLACE EQUATION IN THREE DIMENSIONS

Fundamental Solution

A fundamental solution of Laplace's equation satisfies

$$\Delta u = u_{xx} + u_{yy} + u_{zz} = -\delta(x-x', y-y', z-z'),$$

where the Dirac delta function δ denotes a unit source concentrated at the point (x', y', z'). No function has this property, but it can be thought of as a limit of functions whose integrals over space are unity, and whose support (the region where the function is non-zero) shrinks to a point. The definition of the fundamental solution thus implies that, if the Laplacian of u is integrated over any volume that encloses the source point, then

$$\iiint_V div\nabla u dV = -1.$$

the Laplace equation is unchanged under a rotation of coordinates, and hence we can expect that a fundamental solution may be obtained among solutions that only depend upon the distance r from the source point. If we choose the volume to be a ball of radius a around the source point, then Gauss' divergence theorem implies that;

$$-1 = \iiint_V div\nabla u dV = \iint_S u_r dS = 4\pi a^2 u_r(a).$$

It follows that

$$u_r(r) = -\frac{1}{4\pi r^2},$$

on a sphere of radius r that is centered around the source point, and hence

$$u = \frac{1}{4\pi r}.$$

a similar argument shows that in two dimensions

$$u = \frac{-\log r}{4\pi}.$$

Green's Function

A Green's function is a fundamental solution that also satisfies a suitable condition on the boundary S of a volume V. For instance, $G(x, y, z; x', y', z')$ may satisfy

$$\nabla.\nabla G = -\delta(x - x', y - y', z - z') \text{ in } V,$$

$$G = 0 \text{ if } (x, y, z) \text{ on } S.$$

Now if u is any solution of the Poisson equation in V:

$$\nabla.\nabla\, u = -f,$$

and u assumes the boundary values g on S, then we may apply Green's identity, (a consequence of the divergence theorem) which states that

$$\iiint_V [G\nabla.\nabla u - u\nabla.\nabla G] dV = \iiint_V \nabla.[G\nabla u - u\nabla G] dV$$

$$= \iiint_S [Gu_n - uG_n] dS.$$

The notations u_n and G_n denote normal derivatives on S. In view of the conditions satisfied by u and G, this result simplifies to;

$$u(x', y', z') = \iiint_V Gf dV + \iint_S G_n g dS.$$

Thus the Green's function describes the influence at of the data f and g. For the case of the interior of a sphere of radius a, the Green's function may be obtained by means of a reflection the source point P at distance ñ from the centre of the sphere is reflected along its radial line to a point P' that is at a distance

$$\rho' = \frac{a^2}{\rho}.$$

Note that if P is inside the sphere, then P' will be outside the sphere. The Green's function is then given by

$$\frac{1}{4\pi R} - \frac{a}{4\pi\rho R'},$$

where R denotes the distance to the source point P and R' denotes the distance to the reflected point P'. A consequence of this expression for the Green's function is the Poisson integral formula. Let ρ, θ, and φ be spherical coordinates for the source point P. Here θ denotes the angle with the vertical axis, which is contrary to the usual American mathematical notation, but agrees with standard European and physical practice. Then the solution of the Laplace equation inside the sphere is given by

$$u(P) = \frac{1}{4\pi} a^3 \left(1 - \frac{\rho^2}{a^2}\right) \iint \frac{g(\theta', \varphi') \sin\varphi' d\theta' d\varphi'}{(a^2 + \rho^2 - 2a\rho\cos\Theta)^{3/2}},$$

where

$$\cos\Theta = \cos\varphi\cos\varphi' + \sin\varphi\sin\varphi'\cos(\theta - \theta').$$

A simple consequence of this formula is that if u is a harmonic function, then the value of u at the centre of the sphere is the mean value of its values on the sphere. This mean value property immediately implies that a non-constant harmonic function cannot assume its maximum value at an interior point.

Electrostatic Potential

Electrostatics is the part of physics that describes interactions between stationary charges. You are probably familiar with Coulomb's Law, the central law of electrostatics. This law says that two charged particles exert a force on each other equal to:

$$F = q_1 q_2 / r_{12}^2$$

The electrostatic force, F, is proportional to the product of the charges on the two particles, q_1 and q_2, and inversely proportional to the *square* of the distance separating the particles, r_{12}. Another important characteristic of a charged system is its potential energy, PE. Potential energy is created by electrostatic interactions between charge particles and is equal to:

$$PE = q_1 q_2 / r_{12}$$

Notice that this formula looks nearly the same as Coulomb's Law. The only difference is that potential energy is inversely proportional to the distance between charges, while the Coulomb force is inversely proportional to the *square* of the distance.The most useful quantity for our purposes is the *electrostatic potential.* This quantity is related to PE as follows: the electrostatic potential created by a system of charges at a particular point in space, (x, y, z), is equal to the change in potential energy that occurs when a +1 ion is introduced at this point.

This definition can be made clearer with the aid of the following pictures. Imagine a molecule consisting of an electron density cloud and several positively charged nuclei. Now suppose we want to know the electrostatic potential this molecule creates at point (x, y, z). We can obtain the potential by introducing a +1 charge at (x, y, z) and calculating the change in energy. The new picture looks like this:

The change in energy is simply the potential energy created by interactions between the +1 charge and the charges in the molecule. We can calculate this energy by calculating $(+1)(q_{molecule})/r$ for each charge in the molecule, $q_{molecule}$, and adding up all of these energies. This energy is the molecule's electrostatic potential.

Electrostatic potential is both a molecular property and a spatial property. It depends on what charges exist in the molecule and how they there are distributed. It also depends on what point (x, y, z) we choose to investigate. If we select a point where the +1 charge is attracted by the molecule, the potential will be negative at this point. On the other hand, if we select a point where the +1 charge is repelled, the potential will be positive. Molecules contain many charged particles, nuclei and electrons, and the net impact of these particles on the +1 "probe" can only be determined by a computer. However, since we know that potential energy and distance are inversely related, it is likely that the molecular charge(s) closest to the +1 particle have the largest effect.

For example, the following diagram shows an ionic compound consisting of three ions. It is likely that the potential in the immediate vicinity of each ion is determined largely by this ion, and the more distant ions have relatively small effect. On the other hand, the potential in any region that is near two or more ions must be determined by a careful calculation.This kind of behaviour is seen in practically every system. Positive particles, like atomic nuclei or polyatomic cations, are surrounded by regions of positive potential.

Likewise, negative particles, like polyatomic anions, are surrounded by regions of negative potential. In a few pages I will show you how to use electrostatic potentials to make qualitative statements about atomic charges. When I do this, I will assume that the potential in a given region is controlled by the "local" atom, so a positive potential will indicate a positively charged "local" atom and a negative potential will indicate a negatively charged "local" atom.

Electrostatic Potential Maps

The same Spartan 02 computer programme that is used to construct molecular models, and calculate model electron density clouds, is also be used to calculate model electrostatic potentials. The process is complicated because the programme must calculate the interaction between the +1 probe charge and every part of the electron density cloud, and it is necessary to divide the cloud into many small pieces to do this accurately.

The display of electrostatic potential data is another difficult problem. We cannot look at the potential everywhere simultaneously, so we must select some points of interest first, and display the potential at these points.

Spartan02 selects the points located on the molecule's size

density surface (0.002 au isodensity surface) and calculates the potential at these points. Then Spartan02 assigns colors to these points according to their potential. The resulting rainbow-colored surface is called a *potential map.*

Default Colour Scale

Spartan02 uses colour to show the value of the potential at different points on the map. Each potential corresponds to a unique colour, and you "read" a map by noticing where different colors appear.

Spartan02 can select colors automatically or it can select colors according to instructions that you give it. When Spartan02 is allowed to select colors automatically, it is uses a *default colour scale* to equate different potentials with different colors.

First, the programme surveys all of the potentials on the map, and finds the most negative and most positive potentials. These regions are colored red and dark blue, respectively. Intermediate potentials are then assigned colors according to the standard colour spectrum:

red (most negative) < yellow < green < light blue
< dark blue (most positive)

The potentials on this map range from –46 to +48 kcal/mol. The red region near oxygen shows the location of the most negative potential, –46 kcal/mol (this region attracts the +1 probe). The dark blue regions near the hydrogens show the locations of the most positive potential, +48 kcal/mol (these regions repel the +1 probe).

The green region shows the location of the mean potential, the average of the two extremes. On this map, the mean potential is +1 kcal/mol. The yellow region and light blue regions show the locations of potentials halfway between the mean (green) and extremes (red and dark blue). On this map, yellow corresponds to –22 kcal/mol and light blue corresponds to +24 kcal/mol.

Standard Colour Scale

When we use the default colour scale, we know that the

red and dark blue regions correspond to the *extreme* values of the electrostatic potential, but how can we know what these extreme potentials are? Different maps have different extremes. As a result, if we don't know the numerical value of the extreme potentials, we might be fooled into thinking that we are looking at large extremes (and large differences between extremes) when we are not.

Consider the potential map of propane, $CH_3CH_2CH_3$, shown below. This map is colored according to the default scale. It turns blue near every hydrogen, and yellow-orange around every carbon (the "cavity" at the top of the map is bordered by all three carbons and is red). These colors seem to suggest that the potential swings wildly between hydrogen (medium-dark blue) and carbon (yellow-red).

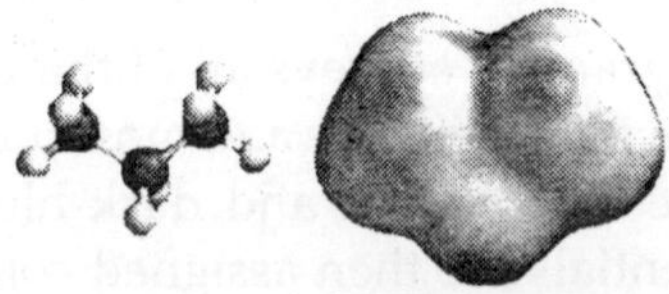

Fig. Potential Map of Propane, $CH_3CH_2CH_3$

Now examine the potential map of 1-propanol, $CH_3CH_2CH_2OH$, shown below. This map is also colored according to its default colour scale. Extreme potentials are located near the OH group. The $CH_3CH_2CH_2$ group is almost uniformly green. This suggests that the potential does not change much between hydrogen and carbon in this molecule.

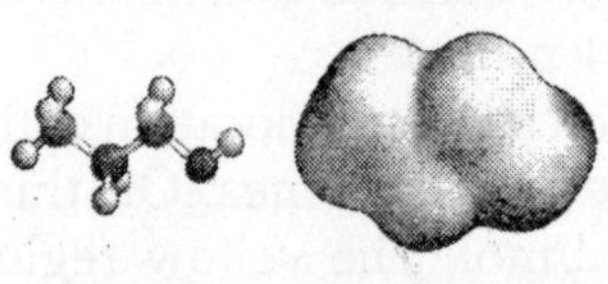

Fig. Potential Map of 1-propanol, $CH_3CH_2CH_2OH$

There does not appear to be any consistency in these maps. The $CH_3CH_2CH_2$ group appears to generate extreme potentials in propane, but a nearly constant potential in 1-propanol. If we carry this idea to its logical conclusion, we would say that the $CH_3CH_2CH_2$ groups are radically different in these molecules. Unfortunately, this conclusion is contradicted by several other types of data (bond distances, isodensity surfaces, electronegativity data, and so on), so we must seek another explanation. The answer lies in the numerical value of the extreme potentials. The potentials on the propane map swing from –22 to +4 kcal/mol, while the potentials on the 1-propanol map cover a much wider range, –48 to +44 kcal/mol. Since both maps use the same number of colors, each colour on 1–propanol's map covers a wider range of potentials. This is why the entire $CH_3CH_2CH_2$ group has roughly the same colour on the 1–propanol map.

We can avoid these problems by giving the values of the extreme potentials for every map. When the extreme potentials are known, the meaning of other colors can be inferred as follows:

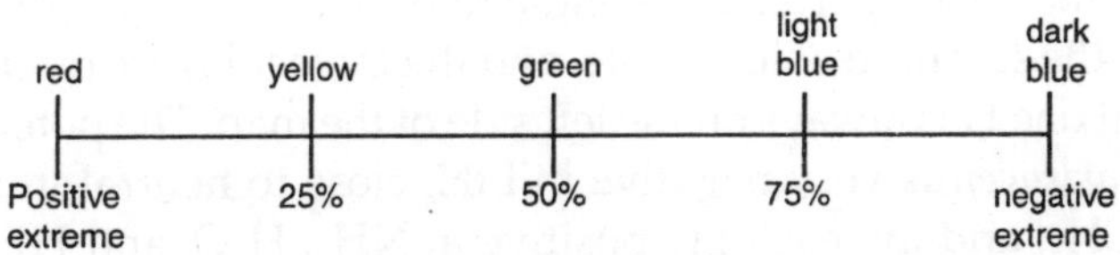

Green always corresponds to a potential halfway between the two extremes (red/dark blue). Yellow and light blue split the difference between the mean (green) and the extremes (red/dark blue). Applying this to the 1-propanol map (red = –48 and blue = +44 kcal/mol) gives green = –2, yellow = –25, and light blue = +21 kcal/mol. Notice that the entire potential range of propane's map, –22 to +4 kcal/mol, would be assigned two colors, green and yellow, on 1-propanol's map.

Another solution that we can also use is to colour the maps of *neutral* molecules using a *standard colour scale.* My definition of this scale places the extremes at –40 (red) and +40 kcal/mol (dark blue). This means that intermediate colors correspond to modest potentials of –20 (yellow), 0 (green), and +20 (light blue) kcal/mol.

The potential maps of propane and 1-propanol colored according to my standard colour scale (remember that we have not changed the potentials, just the colors assigned to each potential). Now the $CH_3CH_2CH_2$ group looks roughly the same in both molecules.

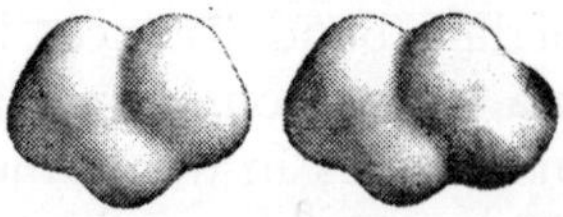

Fig. Potential Maps of Propane, $CH_3CH_2CH_3$, and 1-propanol, $CH_3CH_2CH_2OH$ (Standard Colour Scale)

The standard colour scale is useful, but it is not entirely free of problems. Suppose the potentials on a given map lie *outside* the standard extremes, that is, suppose some potentials are less than –40 or greater than +40 kcal/mol.

POLAR H-X BONDS

The isodensity surfaces can be used to get information about H-X bond polarity. This information can be obtained much more easily from a potential map. The maps are colored using the standard colour scale, and the molecules are oriented so that one H is always on the left side of the map. The potential *near hydrogen* is very negative in LiH, close to neutral in BH_3 and CH_4, and increasingly positive in NH_3, H_2O, and HF. As we expect the potentials closely track the charge on hydrogen.

Potential maps (standard colour scale) of LiH, BH_3, CH_4, NH_3, H_2O, HF

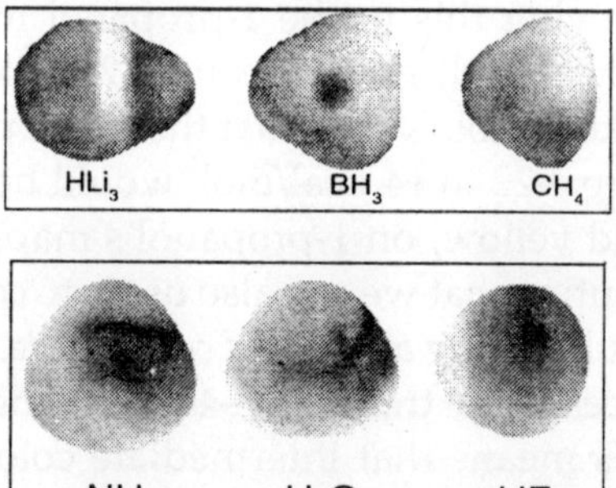

Although the potential near hydrogen varies systematically with the electronegativity of X, the potential near X does not. N and O are surrounded by more negative potentials than F, an unexpected result since F is more electronegative.

This strange behaviour is due to a combination of factors,

one of which is the non-spherical shape of X. It is not hard to see how a non-spherical electron density cloud might create unusual electrostatic potentials. Consider the cartoon atoms shown below. If the nucleus is at the centre of a spherical cloud (pink region), the potential is zero everywhere outside the atom (A). However, if the nucleus is set to one side of the cloud (B), or the cloud is non-spherical (C), unusual potentials are created.

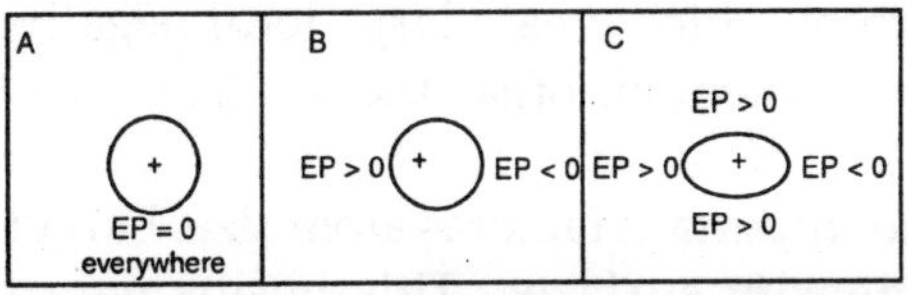

Chemists refer to the regions of positive and negative potential as *electron-poor* and *electron-rich,* respectively. Thus, the red regions in the NH_3 and H_2O maps appear to be electron-rich.

POLAR C–X BONDS

Carbon is only slightly more electronegative than hydrogen. Therefore, we expect C–F, C–O, and C–N bonds to be polar, just like the corresponding H–F, H–O, and H–N bonds. Unfortunately, polar bonds involving carbon are almost impossible to detect using potential maps.

Methyl propionate, $CH_3CH_2C(=O)OCH_3$, furnishes us with a good example of this phenomenon. Two carbons in this molecule are bonded to oxygen. The *methyl* carbon makes one CO bond, O–CH_3, and the *carbonyl* carbon makes three CO bonds, C(=O)–O.

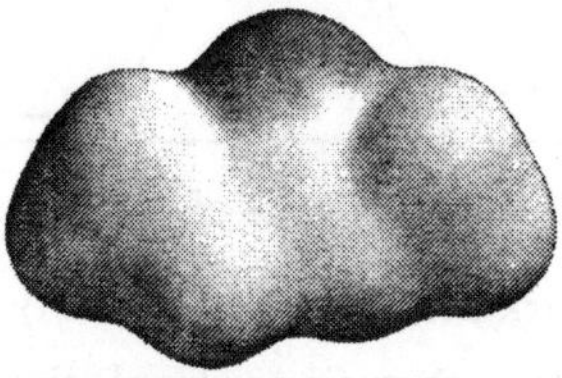

Fig. Potential Map of Methyl Propionate, $CH_3CH_2C(=O)OCH_3$

Since all of the CO bonds are polar, all of them transfer electron density from carbon to oxygen and we expect to find large positive potentials around these carbons. The following potential map of methyl propionate standard colour scale) shows that this does not happen. The potential near each carbon is very small.

Two factors account for these observations. First, the "local" atom principle breaks down near carbon whenever it is sandwiched between several other atoms. Notice that the carbonyl carbon in methyl propionate lies in between two oxygens, both of which create large local negative potentials. These negative potentials offset the positive potentials created by carbon.

A second important factor is atom size. Larger atoms have larger size density surfaces. This means we calculate and display potential at regions farther away from the atom. Since potential falls off as $1/r$, a potential map will show smaller potentials near a large atom than it will near a small atom, other factors being equal.

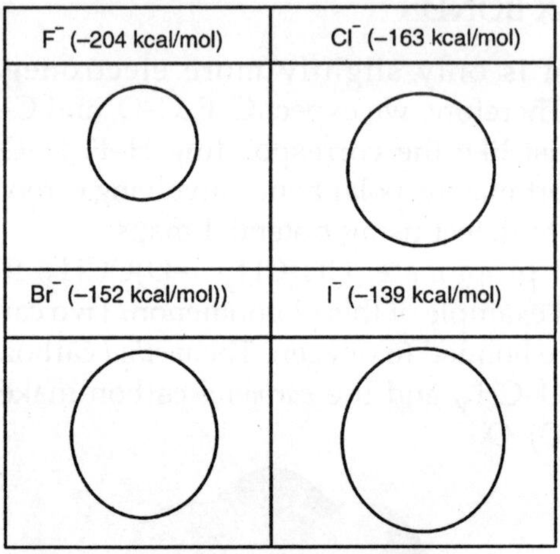

Fig. Potential maps of halide anions

(red = –200, dark blue = –120 kcal/mol): F^-, Cl^-, Br^-, I^-.

All of these ions carry identical charges of –1, but the potential is most negative around the smallest anion, fluoride. As ion and map size increase, the map's potential becomes steadily less negative.

CHARGE DISTRIBUTION IN IONS

Electrostatic potential maps can be used to analyse charge distributions in ions. Although an ion's Lewis structure necessarily contains a (formally) charged atom, the actual distribution of charge may be quite different.

To begin, consider the potential maps of CH_3^- and HO^- shown below (colour scale = –200 to –120 kcal/mol). The Lewis structures of these ions assign a –1 formal charge to C and O, respectively. The potential maps support these charge assignments in that the most negative potentials appear near the charged atoms. Lewis structures appear to give accurate charge information for these ions.

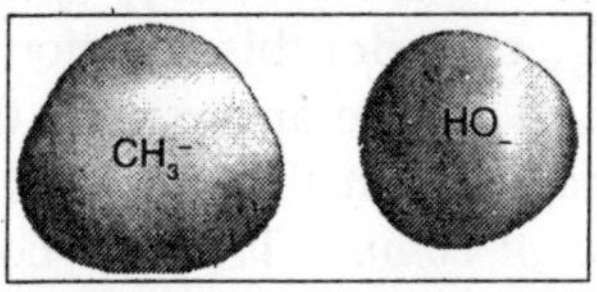

Fig. Potential Maps of CH_3^- and HO^-

(red = –200, dark blue = –120 kcal/mol)

Next, consider the potential maps of NH_4^+ and H_3O^+ shown below (colour scale = +120 to +200 kcal/mol). The Lewis structures of these ions assign a +1 formal charge to N and O, respectively, but the potential maps do not support these charge assignments. According to the maps, the most positive potentials appear near the hydrogens. We must conclude that the hydrogens carry the positive charge, and the charge is spread equally over all of the hydrogens. Lewis structures do not give accurate charge information for these ions.

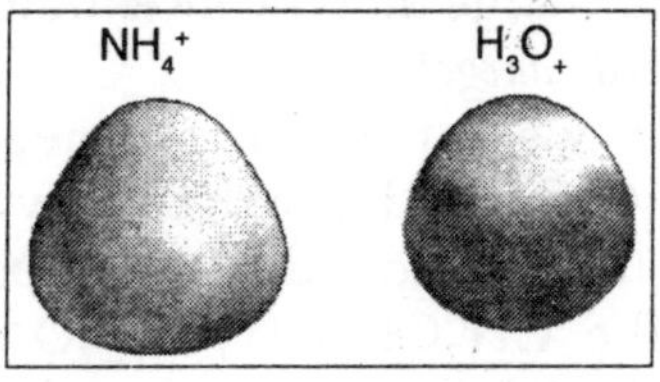

Fig. Potential Maps of NH_4^+ and H_3O^+

(red = +120, dark blue = +200 kcal/mol)

Why do Lewis structures fail to get the charges right in these cations? The answer to this in the assumption underlying formal charge assignments: bonded atoms share electrons equally. The O in H_3O^+ carries *a* + 1 charge if it shares bonding electrons equally with the neighboring H. We know, however, that O is more electronegative than H (and O^+ is more electronegative still), so this assumption cannot be correct. Formal charges can give the wrong picture.

So far we have focused on charge location, the red region on an anion's map and the blue region on a cation's map. What about the other colors/regions on these maps? Does the red region near oxygen in H_3O^+ indicate that oxygen is negative?

Before you take the bait and say, "yes, a red O must be negatively charged", consider this: a cation repels a +1 charge no matter how the +1 charge approaches the cation. In other words, all of the regions on the potential map of H_3O^+ are positive *regardless of colour.* The red colour on the map corresponds to a positive potential, as do all of the other colors. We cannot guess the charges of atoms near any of these other colors. Potential maps are especially valuable tools for identifying and characterizing ionic resonance hybrids. Partial bonding in an ionic resonance hybrid can spread the ion's charge over two or more atoms. This type of charge is known as a *delocalized charge.* Each charged atom carries only a portion of the delocalized charge, and these atoms create noticeably smaller "local" potentials than atoms with full charges.

One way to detect delocalized charge is to compare the potential map of a suspected ionic resonance hybrid with the map of an ion whose charge is concentrated on a single atom of the same type (a charge that is concentrated on one atom is known as a *localized charge*).

For example, we can tell if the negative charge in $CH_3CH_2CH_2C(=O)O^-$ is delocalized by comparing the potential map of this compound with the potential map of a localized ion like $CH_3CH_2CH_2CH_{-2}O^-$. To make this comparison meaningful, we use the same colour scale for both maps, and we compare charged atoms (oxygens) of similar size.

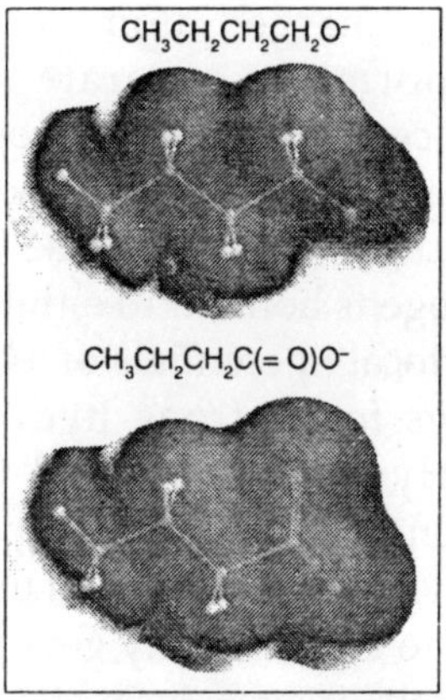

Fig. Potential Maps for Oxygen Anions
(Colour Scale = –174 to –55 kcal/mol)

The two oxygen atoms in the suspect compound produce significantly smaller negative potentials (yellow-orange) than the oxygen atom in the reference ion (red).

In addition, both oxygens in the suspect compound produce similar potentials. These observations suggest that the charge in $CH_3CH_2CH_2C(=O)O^-$ is delocalized over both oxygens:

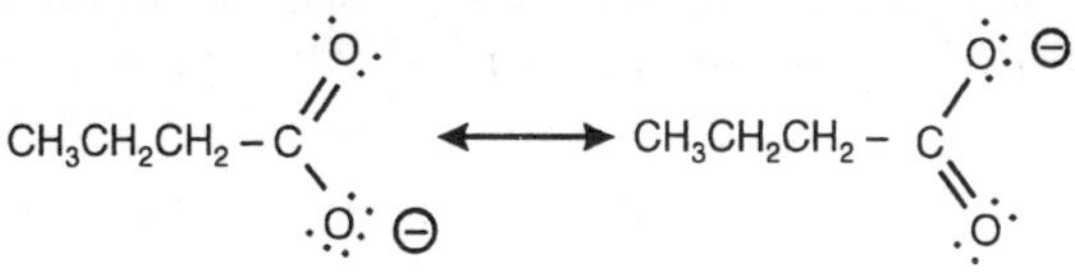

HYDROGEN BONDS

One of the most important nonbonded interactions that can be studied with potential maps is the *hydrogen bond* or *H bond.* This kind of "bond" occurs when two atoms, a positively charged hydrogen atom and a negatively charged non-hydrogen atom, interact electrostatically.

Hydrogen bonds occur most often between neutral molecules. The interacting atoms in these molecules carry fractional charges, and this type of hydrogen bond can be considered a type of dipole-dipole interaction.

The fractionally charged atoms responsible for most hydrogen bonds normally generate significant "local" potentials. Experience shows that hydrogen bonds typically involve a hydrogen atom whose "local" potential is ≥+ 40 kcal/mol, and a non-hydrogen partner whose "local" potential is ≤ –40 kcal/mol. Hydrogens bonded to either oxygen or fluorine typically generate "local" potentials of +40 kcal/mol or more. These atoms always form strong hydrogen bonds with a suitable partner. Hydrogens bonded to certain types of nitrogen can also generate large "local" potentials, and many hydrogen-bonded systems are based on this type of hydrogen. Hydrogens bonded to carbon rarely generate large potentials, and CH_n groups rarely, if ever, form strong hydrogen bonds.

On the other side of the ledger, only nitrogen and oxygen atoms seem to generate large negative "local" potentials in neutral compounds. Therefore, most hydrogen bonds look like O–H... X or N–H... X, where X = N or O. An example of the O–H... O pattern can be found in a cup of water. The two oxygen atoms belong to different water molecules, and we can draw the hydrogen bonded water molecules as HO–H... OH_{-2}.

Several examples of the N–H... X pattern appear in double-stranded DNA. Each DNA strand contains four types of rings (called nucleotides) that are responsible for forming hydrogen bonds with rings in the other strand. The hydrogen bonding capacity of each nucleotide also controls DNA synthesis, and insures that a newly synthesized strand contains the proper sequence of nucleotides.

Two nucleotides, cytosine and guanine, can form three hydrogen bonds simultaneously. The other nucleotides, adenine and thymine, can form two. All five hydrogen bonds follow the N–H... X pattern, and potential maps show that the potentials near these hydrogens are all greater than +40 kcal/mol. Potential maps of cytosine and guanine are shown below. The molecules have been pushed apart so that their respective potential maps can be compared. The maps show that all three of the N–H groups responsible for hydrogen bonds generate large "local" positive potentials. The maps also show that the nitrogen and oxygen partners generate large "local" negative potentials. Each molecule generates a complementary pattern of positive and negative potentials so that, when they approach, three stabilizing interactions will be created.

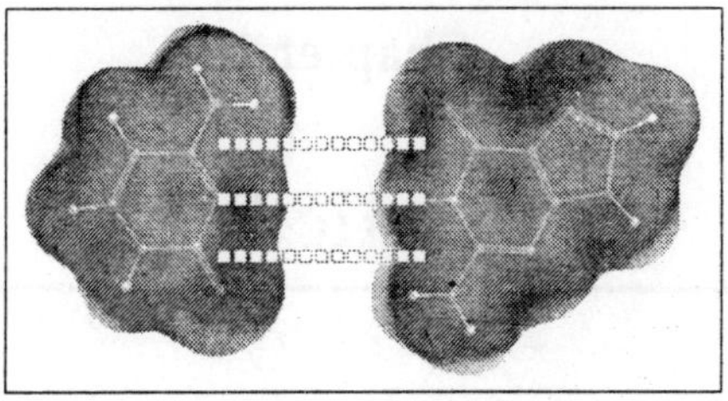

Fig. Hydrogen Bond Partners are Identified by Dashed Lines

Chapter 6

Magnetic Force

Magnets are very common items in the workplace and household. Uses of magnets range from holding pictures on the refrigerator to causing torque in electric motors. Most people are familiar with the general properties of magnets but are less familiar with the source of magnetism. The traditional concept of magnetism centers around the magnetic field and what is know as a dipole.

The term "magnetic field" simply describes a volume of space where there is a change in energy within that volume. This change in energy can be detected and measured. The location where a magnetic field can be detected exiting or entering a material is called a magnetic pole. Magnetic poles have never been detected in isolation but always occur in pairs, hence the name dipole. Therefore, a dipole is an object that has a magnetic pole on one end and a second, equal but opposite, magnetic pole on the other.

A bar magnet can be considered a dipole with a north pole at one end and south pole at the other. A magnetic field can be measured leaving the dipole at the north pole and returning the magnet at the south pole. If a magnet is cut in two, two magnets or dipoles are created out of one. This sectioning and creation of dipoles can continue to the atomic level. Therefore, the source of magnetism lies in the basic building block of all matter the atom.

Magnetism is a force generated in matter by the motion of electrons within its atoms. Magnetism and electricity represent different aspects of the force of electromagnetism, which is one part of Nature's fundamental electroweak force.

The region in space that is penetrated by the imaginary lines of magnetic force describes a magnetic field. The strength of the magnetic field is determined by the number of lines of force per unit area of space.

Magnetic fields are created on a large scale either by the passage of an electric current through magnetic metals or by magnetized materials called magnets. The elemental metals-iron, cobalt, nickel, and their solid solutions or alloys with related metallic elements-are typical materials that respond strongly to magnetic fields. Unlike the all-pervasive fundamental force field of gravity, the magnetic force field within a magnetized body, such as a bar magnet, is polarized-that is, the field is strongest and of opposite signs at the two extremities or poles of the magnet.

THE SOURCE OF MAGNETISM

All matter is composed of atoms, and atoms are composed of protons, neutrons and electrons. The protons and neutrons are located in the atom's nucleus and the electrons are in constant motion around the nucleus. Electrons carry a negative electrical charge and produce a magnetic field as they move through space. A magnetic field is produced whenever an electrical charge is in motion. The strength of this field is called the magnetic moment.

This may be hard to visualize on a subatomic scale but consider electric current flowing through a conductor. When the electrons (electric current) are flowing through the conductor, a magnetic field forms around the conductor. The magnetic field can be detected using a compass. The magnetic field will place a force on the compass needle, which is another example of a dipole.Since all matter is comprised of atoms, all materials are affected in some way by a magnetic field. However, not all materials react the same way.

HISTORY OF MAGNETISM

The history of magnetism dates back to earlier than 600 b.c., but it is only in the twentieth century that scientists have begun to understand it, and develop technologies based on this understanding. Magnetism was most probably first observed in a form of the mineral magnetite called lodestone, which consists of iron oxide-a chemical compound of iron and oxygen. The ancient Greeks were the first known to have used this mineral, which they called a magnet because of its ability to attract other pieces of the same material and iron.

The Englishman William Gilbert (1540-1603) was the first to investigate the phenomenon of magnetism systematically using scientific methods. He also discovered that the Earth is itself a weak magnet. Early theoretical investigations into the nature of the Earth's magnetism were carried out by the German Carl Friedrich Gauss (1777-1855). Quantitative studies of magnetic phenomena initiated in the eighteenth century by Frenchman Charles Coulomb (1736-1806), who established the inverse square law of force, which states that the attractive force between two magnetized objects is directly proportional to the product of their individual fields and inversely proportional to the square of the distance between them.

Danish physicist Hans Christian Oersted(1777-1851) first suggested a link between electricity and magnetism. Experiments involving the effects of magnetic and electric fields on one another were then conducted by Frenchman Andre Marie Ampere (1775-1836) and Englishman Michael Faraday (1791-1869), but it was the Scotsman, James Clerk Maxwell (1831-1879), who provided the theoretical foundation to the physics of electromagnetism in the nineteenth century by showing that electricity and magnetism represent different aspects of the same fundamental force field. Then, in the late 1960s American Steven Weinberg and Pakistani Abdus Salam, performed yet another act of theoretical synthesis of the fundamental forces by showing that electromagnetism is one part of the electroweak force.

The modern understanding of magnetic phenomena in condensed matter originates from the work of two Frenchmen: Pierre Curie (1859-1906), the husband and scientific collaborator of Madame Marie Curie (1867-1934), and Pierre Weiss (1865-1940). Curie examined the effect of temperature on magnetic materials and observed that magnetism disappeared suddenly above a certain critical temperature in materials like iron. Weiss proposed a theory of magnetism based on an internal molecular field proportional to the average magnetization that spontaneously align the electronic micromagnets in magnetic matter.

The present day understanding of magnetism based on the theory of the motion and interactions of electrons in atoms (called quantum electrodynamics) stems from the work and theoretical models of two Germans, Ernest Ising (1900-) and Werner Heisenberg (1901-1976). Werner Heisenberg was also one of the founding fathers of modern quantum mechanics. Magnetism arises from two types of motions of electrons in atoms-one is the motion of the electrons in an orbit around the nucleus, similar to the motion of the planets in our solar system around the sun, and the other is the spin of the electrons around its axis, analogous to the rotation of the Earth about its own axis.

The orbital and the spin motion independently impart a magnetic moment on each electron causing each of them to behave as a tiny magnet. The magnetic moment of a magnet is defined by the rotational force experienced by it in a magnetic field of unit strength acting perpendicular to its magnetic axis. In a large fraction of the elements, the magnetic moment of the electrons cancel out because of the Pauli exclusion principle, which states that each electronic orbit can be occupied by only two electrons of opposite spin. However, a number of so-called transition metal atoms, such as iron, cobalt, and nickel, have magnetic moments that are not cancelled; these elements are, therefore, common examples of magnetic materials. In these transition metal elements the magnetic moment arises only from the spin of the electrons.

In the rare earth elements (that begin with lanthanum in the sixth row of the Periodic Table of Elements), however, the effect of the orbital motion of the electrons is not cancelled, and hence both spin and orbital motion contribute to the magnetic moment. Examples of some magnetic rare earth elements are: cerium, neodymium, samarium, and europium. In addition to metals and alloys of transition and rare earth elements, magnetic moments are also observed in a wide variety of chemical compounds involving these elements. Among the common magnetic compounds are the metal oxides, which are chemically bonded compositions of metals with oxygen.

The Earth's geomagnetic field is the result of electric currents produced by the slow convective motion of its liquid core in accordance with a basic law of electromagnetism which states that a magnetic field is generated by the passage of an electric current. According to this model, the Earth's core should be electrically conductive enough to allow generation and transport of an electric current. The geomagnetic field generated will be dipolar in character, similar to the magnetic field in a conventional magnet, with lines of magnetic force lying in approximate planes passing through the geomagnetic axis. The principle of the compass needle used by the ancient mariners involves the alignment of a magnetized needle along the Earth's magnetic axis with the imaginary south pole of the needle pointing towards the magnetic north pole of the Earth. The magnetic north pole of the Earth is inclined at an angle of 11 degrees away from its geographical north pole.

TYPES OF MAGNETISM

Five basic types of magnetism have been observed and classified on the basis of the magnetic behaviour of materials in response to magnetic fields at different temperatures. These types of magnetism are: ferromagnetism, ferrimagnetism, antiferromagnetism, paramagnetism, and diamagnetism.

Ferromagnetism and ferrimagnetism occur when the magnetic moments in a magnetic material line up spontaneously at a temperature below the so-called Curie temperature, to produce net magnetization. The magnetic moments are aligned at random at temperatures above the Curie point, but become ordered, typically in a vertical or, in special cases, in a spiral (helical) array, below this temperature. In a ferromagnet magnetic moments of equal magnitude arrange themselves in parallel to each other. In a ferrimagnet, on the other hand, the moments are unequal in magnitude and order in an antiparallel arrangement. When the moments are equal in magnitude and ordering occurs at a temperature called the Neel temperature in an antiparallel array to give no net magnetization, the phenomenon is referred to as antiferromagnetism. These transitions from disorder to order represent classic examples of phase transitions. Another example of a phase transition is the freezing of the disordered molecules of water at a critical temperature of 32°F (0°C) to form the ordered structure of ice.

The magnetic moments-referred to as spins-are localized on the tiny electronic magnets within the atoms of the solid. Mathematically, the electronic spins are equal to the angular momentum (the rotational velocity times the moment of inertia) of the rotating electrons. The spins in a ferromagnetic or a ferrimagnetic single crystal undergo spontaneous alignment to form a macroscopic (large scale) magnetized object. Most magnetic solids, however, are not single crystals, but consist of single crystal domains separated by domain walls. The spins align within a domain below the Curie temperature, independently of any external magnetic field, but the domains have to be aligned in a magnetic field in order to produce a macroscopic magnetized object.

This process is effected by the rotation of the direction of the spins in the domain wall under the influence of the magnetic field, resulting in a displacement of the wall and the eventual creation of a single large domain with the same spin orientation.

Paramagnetism is a weak form of magnetism observed in substances which display a positive response to an applied magnetic field. This response is described by its magnetic susceptibility per unit volume, which is a dimensionless quantity defined by the ratio of the magnetic moment to the magnetic field intensity.

Paramagnetism is observed, for example, in atoms and molecules with an odd number of electrons, since here the net magnetic moment cannot be zero. Diamagnetism is associated with materials that have a negative magnetic susceptibility. It occurs in nonmagnetic substances like graphite, copper, silver and gold, and in the superconducting state of certain elemental and compound metals.

The negative magnetic susceptibility in these materials is the result of a current induced in the electron orbits of the atoms by the applied magnetic field. The electron current then induces a magnetic moment of opposite sign to that of the applied field. The net result of these interactions is that the material is shielded from penetration by the applied magnetic field.

MEASUREMENT OF MAGNETIC FIELD

The magnetic field or flux density is measured in metric units of a gauss (G) and the corresponding international system unit of a tesla (T). The magnetic field strength is measured in metric units of oersteds (Oe) and international units of amperes per meter (A/m). Instruments called gaussmeters and magnetometers are used to measure the magnitude of magnetic fields.

One form of the gaussmeter that is used commonly in the laboratory consists of a current carrying semiconducting element called the Hall probe, which is placed perpendicular to the magnetic field being measured. As a consequence of the so-called Hall effect, a voltage perpendicular to the field and to the current is generated in the probe. This induced voltage is proportional to the magnetic field being measured and can be simply measured using a voltmeter.

Magnetometers are extremely sensitive magnetic field detectors. In one commonly used form the magnetic force is detected by means of a sensitive electronic balance. In this instrument the magnetic substance is placed on one arm of a balance, which in turn is placed in a magnetic field. The magnetic force on the sample is then determined by the weight required to balance the force generated by the magnetic field.

The most sensitive magnetometer in a modern physics laboratory utilizes a magnetic sensing element called the SQUID (which stands for Superconducting QUantum Interference Device). A SQUID consists of an extremely thin electrically resistive junction (called a Josephson junction) between two superconductors. Superconductors are materials which undergo a transition at low temperatures to a state of zero electrical resistance and nearly complete exclusion of magnetic fields.

In its direct current mode of operation, a SQUID is first cooled down to its superconducting state, and then a current is passed through it while the voltage across the junction is monitored. When the junction senses a magnetic field, the flow of current is altered due to an interference phenomenon at the quantum level between two electron wave fronts through the junction, resulting in a change in voltage.

Interference is a phenomenon that occurs generally due to the mixing of two wave fronts; the waves add up in some regions and cancel out in others depending on the location of the crest and trough of each wave in space. For example, the interference between the sound waves from two simultaneously played musical instruments tuned at somewhat different frequencies results in the occurrence of beats or modulations in the sound intensity.

A variation of the SQUID magnetometer is the SQUID gradiometer which measures differences in magnetic fields at different positions. Using this type of instrument magnetic field variations in the femtotesla (10^{-15} tesla) range can be detected. Devices of this type have been used to map the tiny magnetic signals from the human brain.

APPLICATIONS OF MAGNETISM

Electromagnets are utilized as key components of transformers in power supplies that convert electrical energy from a wall outlet into direct current energy for a wide range of electronic devices, and in motors and generators. High field superconducting magnets (where superconducting coils generate the magnetic field) provide the magnetic field in MRI (magnetic resonance imaging) devices that are now used extensively in hospitals and medical centers.

Magnetic materials that are difficult to demagnetize are used to construct permanent magnets. Permanent magnet applications are in loudspeakers, earphones, electric meters, and small motors. A loudspeaker consists of a wire carrying an alternating current. When the wire is in the magnetic field of the permanent magnet it experiences a force that generates a sound wave by alternate compression and rarefaction of the surrounding air when the alternating frequency of the current is in the audible range.

The more esoteric applications of magnetism are in the area of magnetic recording and storage devices in computers, and in audio and video systems. Magnetic storage devices work on the principle of two stable magnetic states represented by the 0 and 1 in the binary number system. Floppy disks have dozens of tracks on which data can be digitally written in or stored by means of a write-head and then accessed or read by means of a read-head.

A write-head provides a strong local magnetic field to the region through which the storage track of the disk is passed. The read-head senses stray magnetic flux from the storage track of the disk as it passes over the head. Another example of digital magnetic storage and reading is the magnetic strip on the back of plastic debit and credit cards. The magnetic strip contains identification data which can be accessed through, for example, an automatic teller machine.

SOME CURRENT RESEARCH TRENDS IN MAGNETISM

Ideally pure magnetic systems have provided the most extensively investigated models of the large scale collective behaviour of atoms and electrons that occur in the vicinity of the critical point of phase transitions. Fascinating effects caused by the intentional introduction of impurities and defects into random locations in the atomic lattice of a magnetic material. For example, these random magnetic systems display transitions to states of order that have no counterparts in pure systems, because pure systems are, by necessity, always close to thermodynamic equilibrium or stability. For these reasons there is now intense interest and research activity in disordered systems, and random magnets provide ideal model systems for such investigations.

An area of intense current activity centers around the search for a likely magnetic pairing force in the high temperature ceramic superconductors that were discovered in 1987 by the German-Swiss team of Georg Bednorz and Karl Alexander Muller. A superconductor achieves a zero resistance state by means of a force field that pairs up the conducting electrons within its atoms. The new ceramic materials are antiferromagnets in their undoped state, but on doping start to superconduct at temperatures that are over 182°F (83°C) warmer than conventional pure metal and alloy superconductors.

The effects of extremely high magnetic fields on the properties of condensed matter continues to be an area of high interest. New research areas, such as the search and study of magnetism in organic matter, and the study of diamagnetism and novel magnetic effects in the recently synthesized nanometer-sized (a nanometer is equal to 10^{-9} meter) carbon tubes, are of increasing interest to physicists and material scientists.

DIAMAGNETIC, PARAMAGNETIC, AND FERROMAGNETIC MATERIALS

When a material is placed within a magnetic field, the magnetic forces of the material's electrons will be affected. This effect is known as Faraday's Law of Magnetic Induction. However, materials can react quite differently to the presence of an external magnetic field. This reaction is dependent on a number of factors, such as the atomic and molecular structure of the material, and the net magnetic field associated with the atoms. The magnetic moments associated with atoms have three origins. These are the electron orbital motion, the change in orbital motion caused by an external magnetic field, and the spin of the electrons.

In most atoms, electrons occur in pairs. Electrons in a pair spin in opposite directions. So, when electrons are paired together, their opposite spins cause their magnetic fields to cancel each other. Therefore, no net magnetic field exists. Alternately, materials with some unpaired electrons will have a net magnetic field and will react more to an external field. Most materials can be classified as diamagnetic, paramagnetic or ferromagnetic.

Diamagnetic metals have a very weak and negative susceptibility to magnetic fields. Diamagnetic materials are slightly repelled by a magnetic field and the material does not retain the magnetic properties when the external field is removed. Diamagnetic materials are solids with all paired electron resulting in no permanent net magnetic moment per atom. Diamagnetic properties arise from the realignment of the electron orbits under the influence of an external magnetic field. Most elements in the periodic table, including copper, silver, and gold, are diamagnetic.

Paramagnetic metals have a small and positive susceptibility to magnetic fields. These materials are slightly attracted by a magnetic field and the material does not retain the magnetic properties when the external field is removed. Paramagnetic properties are due to the presence of some unpaired electrons, and from the realignment of the electron orbits caused by the external magnetic field. Paramagnetic materials include magnesium, molybdenum, lithium, and tantalum.

Ferromagnetic materials have a large and positive susceptibility to an external magnetic field. They exhibit a strong attraction to magnetic fields and are able to retain their magnetic properties after the external field has been removed. Ferromagnetic materials have some unpaired electrons so their atoms have a net magnetic moment. They get their strong magnetic properties due to the presence of magnetic domains. In these domains, large numbers of atom's moments (10^{12} to 10^{15}) are aligned parallel so that the magnetic force within the domain is strong.

When a ferromagnetic material is in the unmagnitized state, the domains are nearly randomly organized and the net magnetic field for the part as a whole is zero. When a magnetizing force is applied, the domains become aligned to produce a strong magnetic field within the part. Iron, nickel, and cobalt are examples of ferromagnetic materials.

Components with these materials are commonly inspected using the magnetic particle method.

MAGNETIC DOMAINS

Ferromagnetic materials get their magnetic properties not only because their atoms carry a magnetic moment but also because the material is made up of small regions known as magnetic domains. In each domain, all of the atomic dipoles are coupled together in a preferential direction. This alignment develops as the material develops its crystalline structure during solidification from the molten state. Magnetic domains can be detected using Magnetic Force Microscopy (MFM) and images of the domains like the one shown below can be constructed.

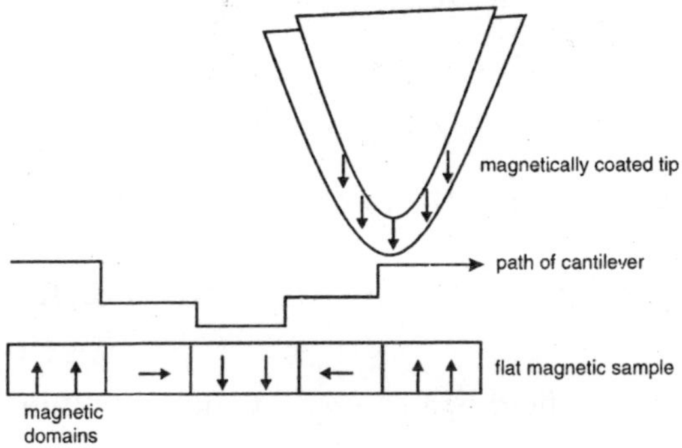

Fig. Magnetic Force Microscopy

During solidification, a trillion or more atom moments are aligned parallel so that the magnetic force within the domain is strong in one direction. Ferromagnetic materials are said to be characterized by "spontaneous magnetization" since they obtain saturation magnetization in each of the domains without an external magnetic field being applied. Even though the domains are magnetically saturated, the bulk material may not show any signs of magnetism because the domains develop themselves and are randomly oriented relative to each other.

Ferromagnetic materials become magnetized when the magnetic domains within the material are aligned. This can be done by placing the material in a strong external magnetic field or by passing electrical current through the material. Some or all of the domains can become aligned. The more domains that are aligned, the stronger the magnetic field in the material. When all of the domains are aligned, the material is said to be magnetically saturated. When a material is magnetically saturated, no additional amount of external magnetization force will cause an increase in its internal level of magnetization.

Fig. Unmagnetized Material

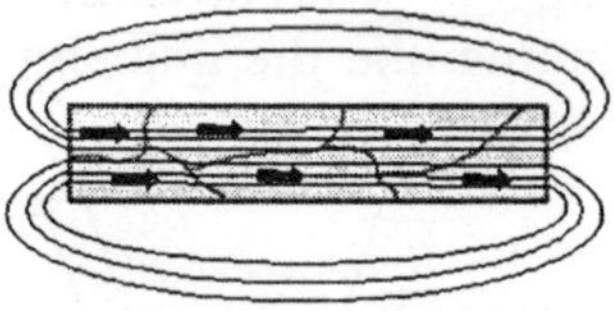

Fig. Magnetized Material

MAGNETIC FIELD CHARACTERISTICS

MAGNETIC FIELD IN AND AROUND A BAR MAGNET

A magnetic field is a change in energy within a volume of space. The magnetic field surrounding a bar magnet can be seen in the magnetograph below. A magnetograph can be created by placing a piece of paper over a magnet and sprinkling the paper with iron filings.

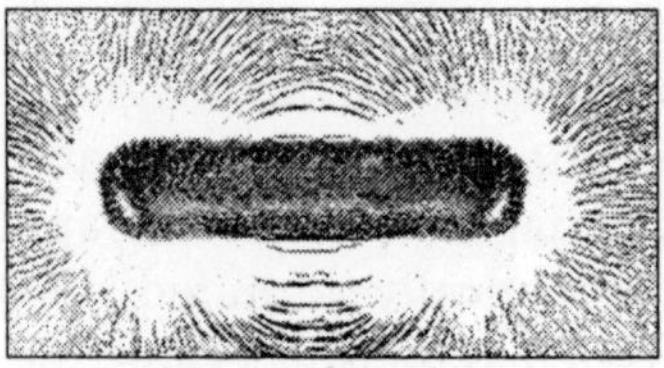

Fig. Bar Magnet

The particles align themselves with the lines of magnetic force produced by the magnet. The magnetic lines of force show where the magnetic field exits the material at one pole and reenters the material at another pole along the length of the magnet. It should be noted that the magnetic lines of force exist in three dimensions but are only seen in two dimensions in the image. It can be seen in the magnetograph that there are poles all along the length of the magnet but that the poles are concentrated at the ends of the magnet. The area where the exit poles are concentrated is called the magnet's north pole and the area where the entrance poles are concentrated is called the magnet's south pole.

MAGNETIC FIELDS IN AND AROUND HORSESHOE AND RING MAGNETS

Magnets come in a variety of shapes and one of the more common is the horseshoe (U) magnet. The horseshoe magnet has north and south poles just like a bar magnet but the magnet is curved so the poles lie in the same plane. The magnetic lines of force flow from pole to pole just like in the bar magnet. However, since the poles are located closer together and a more direct path exists for the lines of flux to travel between the poles, the magnetic field is concentrated between the poles.

Fig. Horseshoe and Ring Magnets

If a bar magnet was placed across the end of a horseshoe magnet or if a magnet was formed in the shape of a ring, the lines of magnetic force would not even need to enter the air. The value of such a magnet where the magnetic field is completely contained with the material probably has limited use. However, it is important to understand that the magnetic field can flow in loop within a material.

General Properties of Magnetic Lines of Force

Magnetic lines of force have a number of important properties, which include:

- They seek the path of least resistance between opposite magnetic poles. In a single bar magnet as shown to the right, they attempt to form closed loops from pole to pole.
- They never cross one another.
- They all have the same strength.
- Their density decreases (they spread out) when they move from an area of higher permeability to an area of lower permeability.
- Their density decreases with increasing distance from the poles.
- They are considered to have direction as if flowing, though no actual movement occurs.
- They flow from the south pole to the north pole within a material and north pole to south pole in air.

When a current carrying conductor is formed into a loop or several loops to form a coil, a magnetic field develops that flows through the centre of the loop or coil along its longitudinal axis and circles back around the outside of the loop or coil. The magnetic field circling each loop of wire combines with the fields from the other loops to produce a concentrated field down the centre of the coil.

The magnetic field is essentially uniform down the length of the coil when it is wound tighter. The strength of a coil's magnetic field increases not only with increasing current but also with each loop that is added to the coil. A long, straight coil of wire is called a solenoid and can be used to generate a nearly uniform magnetic field similar to that of a bar magnet. The concentrated magnetic field inside a coil is very useful in magnetizing ferromagnetic materials for inspection using the magnetic particle testing method. Please be aware that the field outside the coil is weak and is not suitable for magnetizing ferromagnetic materials.

QUANTIFYING MAGNETIC PROPERTIES

However, it is necessary to be able to measure and express quantitatively the various characteristics of magnetism. Unfortunately, a number of unit conventions are used. SI units will be used in this material. The advantage of using SI units is that they are traceable back to an agreed set of four base units–meter, kilogram, second, and Ampere.

Quantity		SI Units (Sommerfeld)	SI Units (Kennelly)	CGS Units (Gaussian)
Field	H	A/m	A/m	oersteds
Flux Density (Magnetic Induction)	B	tesla	tesla	gauss
Flux	ϕ	weber	weber	maxwell
Magnetization	M	A/m	–	erg/Oe-cm³

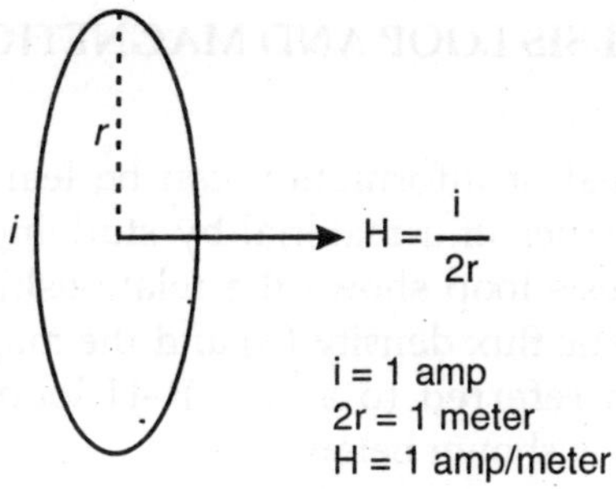

The units for magnetic field strength H are ampere/meter. A magnetic field strength of 1 ampere/meter is produced at the centre of a single circular conductor with a one meter diameter carrying a steady current of 1 ampere.

The number of magnetic lines of force cutting through a plane of a given area at a right angle is known as the magnetic flux density, B. The flux density or magnetic induction has the tesla as its unit. One tesla is equal to 1 Newton/(A/m). From these units, it can be seen that the flux density is a measure of the force applied to a particle by the magnetic field. The Gauss is the CGS unit for flux density and is commonly used by US industry. One gauss represents one line of flux passing through one square centimeter of air oriented 90 degrees to the flux flow.

The total number of lines of magnetic force in a material is called magnetic flux, ϕ. The strength of the flux is determined by the number of magnetic domains that are aligned within a material. The total flux is simply the flux density applied over an area. Flux carries the unit of a weber, which is simply a tesla- meter^2. The magnetization is a measure of the extent to which an object is magnetized. It is a measure of the magnetic dipole moment per unit volume of the object. Magnetization carries the same units as a magnetic field: amperes/meter.

Table. Conversion between CGS and SI Magnetic Units

Quantity	CGS unit	SI units
Magnetic field	$H = 1$ Oe	$H = 1000/4\pi$ A.m^{-1}
Magnetic induction	$B = 1$ gauss	$B = 1 \times 10^{-4}$ Tesla
Magnetization	M = erg.Oe^{-1}.cm^{-3}	$M = 1 \times 103$ A.m^{-1}

THE HYSTERESIS LOOP AND MAGNETIC PROPERTIES

A great deal of information can be learned about the magnetic properties of a material by studying its hysteresis loop. A hysteresis loop shows the relationship between the induced magnetic flux density (*B*) and the magnetizing force (*H*). It is often referred to as the B–H loop. An example hysteresis loop is shown below.

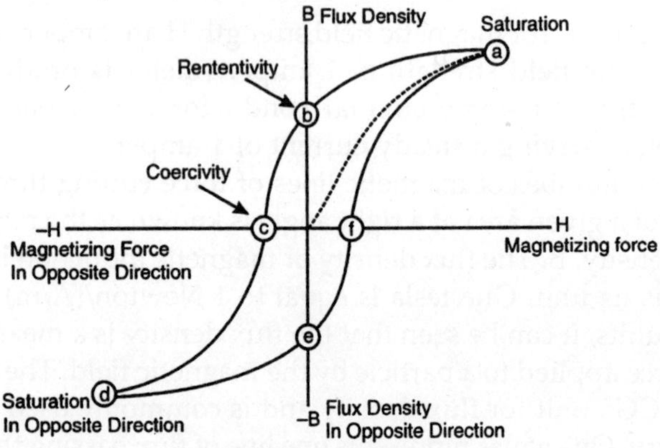

Fig. Hysteresis Loop

The loop is generated by measuring the magnetic flux of a ferromagnetic material while the magnetizing force is changed. A ferromagnetic material that has never been previously magnetized or has been thoroughly demagnetized will follow the dashed line as H is increased. As the line demonstrates, the greater the amount of current applied (H^+), the stronger the magnetic field in the component (B^+). At point "a" almost all of the magnetic domains are aligned and an additional increase in the magnetizing force will produce very little increase in magnetic flux. The material has reached the point of magnetic saturation.

When H is reduced to zero, the curve will move from point "a" to point "b." At this point, it can be seen that some magnetic flux remains in the material even though the magnetizing force is zero. This is referred to as the point of retentivity on the graph and indicates the remanence or level of residual magnetism in the material. As the magnetizing force is reversed, the curve moves to point "c", where the flux has been reduced to zero. This is called the point of coercivity on the curve. The force required to remove the residual magnetism from the material is called the coercive force or coercivity of the material.

As the magnetizing force is increased in the negative direction, the material will again become magnetically saturated but in the opposite direction (point "d"). Reducing H to zero brings the curve to point "e." It will have a level of residual magnetism equal to that achieved in the other direction. Increasing H back in the positive direction will return B to zero. Notice that the curve did not return to the origin of the graph because some force is required to remove the residual magnetism. The curve will take a different path from point "f" back to the saturation point where it with complete the loop.

From the hysteresis loop, a number of primary magnetic properties of a material can be determined.

- *Retentivity*: A measure of the residual flux density corresponding to the saturation induction of a magnetic material. In other words, it is a material's ability to retain a certain amount of residual magnetic field when the magnetizing force is removed after achieving saturation. (The value of B at point b on the hysteresis curve.)

- *Residual Magnetism or Residual Flux*: The magnetic flux density that remains in a material when the magnetizing force is zero. Note that residual magnetism and retentivity are the same when the material has been magnetized to the saturation point. However, the level of residual magnetism may be lower than the retentivity value when the magnetizing force did not reach the saturation level.
- *Coercive Force*: The amount of reverse magnetic field which must be applied to a magnetic material to make the magnetic flux return to zero. (The value of H at point c on the hysteresis curve.)
- *Permeability, μ*: A property of a material that describes the ease with which a magnetic flux is established in the component.
- *Reluctance*: Is the opposition that a ferromagnetic material shows to the establishment of a magnetic field. Reluctance is analogous to the resistance in an electrical circuit.

PERMEABILITY

As previously mentioned, permeability is a material property that describes the ease with which a magnetic flux is established in a component. It is the ratio of the flux density to the magnetizing force and is represented by the following equation:

$$\mu = B/H$$

It is clear that this equation describes the slope of the curve at any point on the hysteresis loop. The permeability value given in papers and reference materials is usually the maximum permeability or the maximum relative permeability. The maximum permeability is the point where the slope of the B/H curve for the unmagnetized material is the greatest. This point is often taken as the point where a straight line from the origin is tangent to the B/H curve.The relative permeability is arrived at by taking the ratio of the material's permeability to the permeability in free space (air).

$$\mu(\text{relative}) = \mu(\text{material}) / \mu(\text{air})$$

Where: $\mu(\text{air}) = 1.256 \times 10^{-6}$ H/m

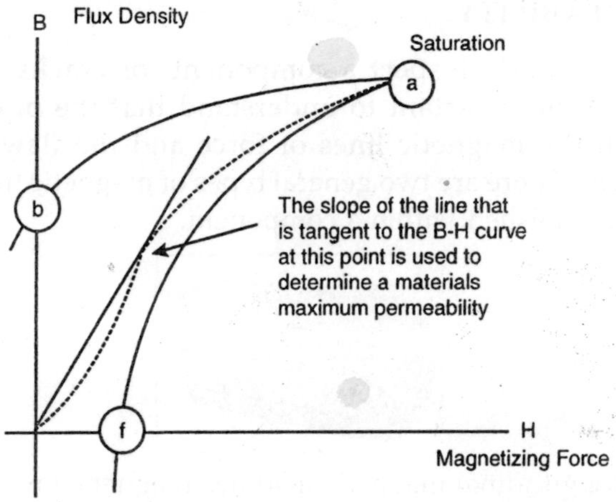

The shape of the hysteresis loop tells a great deal about the material being magnetized. The hysteresis curves of two different materials are shown in the graph.

Relative to other materials, a material with a wider hysteresis loop has:

- Lower Permeability
- Higher Retentivity
- Higher Coercivity
- Higher Reluctance
- Higher Residual Magnetism

Relative to other materials, a material with the narrower hysteresis loop has:

- Higher Permeability
- Lower Retentivity
- Lower Coercivity
- Lower Reluctance
- Lower Residual Magnetism.

In magnetic particle testing, the level of residual magnetism is important. Residual magnetic fields are affected by the permeability, which can be related to the carbon content and alloying of the material. A component with high carbon content will have low permeability and will retain more magnetic flux than a material with low carbon content.

MAGNETIC FIELD ORIENTATION AND FLAW DETECTABILITY

To properly inspect a component for cracks or other defects, it is important to understand that the orientation between the magnetic lines of force and the flaw is very important. There are two general types of magnetic fields that can be established within a component.

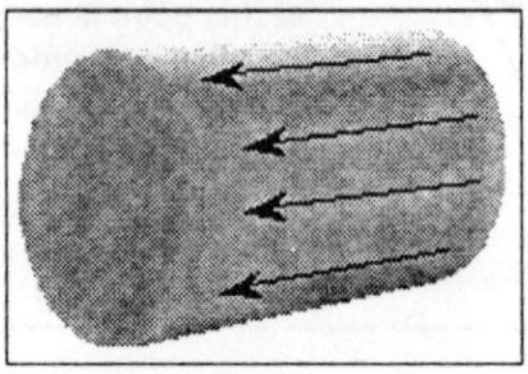

A longitudinal magnetic field has magnetic lines of force that run parallel to the long axis of the part. Longitudinal magnetization of a component can be accomplished using the longitudinal field set up by a coil or solenoid. It can also be accomplished using permanent magnets or electromagnets.

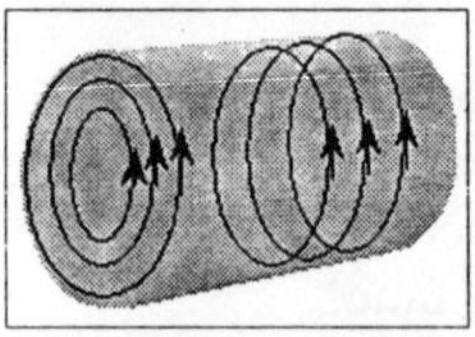

A circular magnetic field has magnetic lines of force that run circumferentially around the perimeter of a part. A circular magnetic field is induced in an article by either passing current through the component or by passing current through a conductor surrounded by the component.

The type of magnetic field established is determined by the method used to magnetize the specimen. Being able to magnetize the part in two directions is important because the best detection of defects occurs when the lines of magnetic force are established at right angles to the longest dimension of the defect.

This orientation creates the largest disruption of the magnetic field within the part and the greatest flux leakage at the surface of the part. An orientation of 45 to 90 degrees between the magnetic field and the defect is necessary to form an indication. Since defects may occur in various and unknown directions, each part is normally magnetized in two directions at right angles to each other.

If the component below is considered, it is known that passing current through the part from end to end will establish a circular magnetic field that will be 90 degrees to the direction of the current. Therefore, defects that have a significant dimension in the direction of the current (longitudinal defects) should be detectable. Alternately, transverse-type defects will not be detectable with circular magnetization.

MAGNETIC FORCES

MAGNETIC FORCE ON A CHARGED PARTICLE

Experimentally, to isolate the nature of the magnetic field, we need to start, as with the electric field, with the simplest thing you can act on, namely a particle. The behaviour of magnetic forces is best indicated by how a charged particle behaves in the presence of such a field. A charged particle moves in the presence of a magnetic field. If the particle stops its motion, the magnetic force on it is zero. This is true independent of the direction or magnitude of the magnetic field. The mathematical description of the force due to the magnetic field is $F = qv \times B$.This formula describes a force which is, by virtue of the cross-product, always perpendicular to v and B and has a magnitude equal to qvB sinq where q is the angle between v and B.

This is a general result for any charged particle moving in any magnetic field. The circular path seen in the applet results because the force is always perpendicular to the velocity (and also to the B field direction). If the velocity magnitude and B are constant, then the path of the particle is a circle since the force, and hence the acceleration, of the particle is always perpendicular to its path and constant in magnitude. This fact is used to make two very practical devices of the 20th century: the mass spectrometer and the cyclotron. Both of these have a multitude of scientific and practical uses.

FORCES ON CURRENTS

In the earliest studies of magnetic fields, individual charged particles were not known. The properties of magnetic fields and forces were studied using the effects of fields on current distributions. If we have a bunch of charges flowing through a straight wire of length L, then the net force is

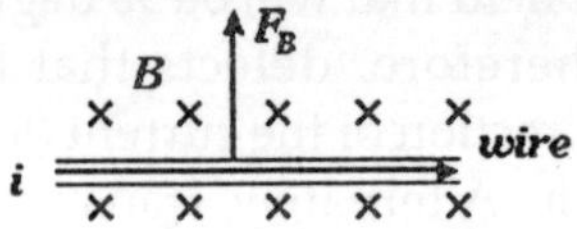

$$\vec{F} = \Sigma_q q\vec{v} \times \vec{B}$$

$$= \Sigma_q \frac{q\vec{L}}{t} \times \vec{B}$$

$$\vec{F} = i\vec{L} \times \vec{B}$$

Force on a wire carrying a current I in a magnetic field B. Where the direction of L is the same as the direction of the current. We can apply the above formulas to the more general case of current distributions of arbitrary shape by breaking any line of current into infinitesimal lengths and integrating over them to find the net force, dF= idL × B

MAGNETIC FORCE ON AN ARBITRARILY SHAPED WIRE

We can generalize the above result to take into account any arbitrary shape of current distribution. For example, consider some arbitrarily shaped wire takes a current i from point a to point b in a magnetic field B which has constant magnitude and direction perpendicular to the plane containing i.

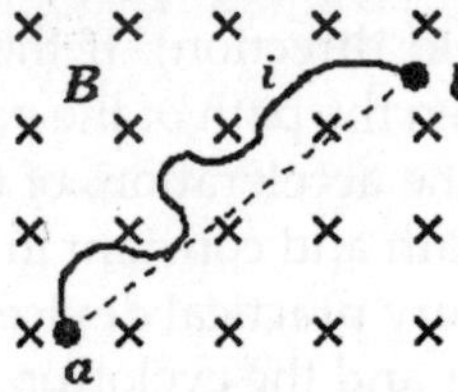

Fig. Current Distribution in an External Magnetic Field

The problem is not as difficult as it might at first appear. Consider any particular piece of the wire.

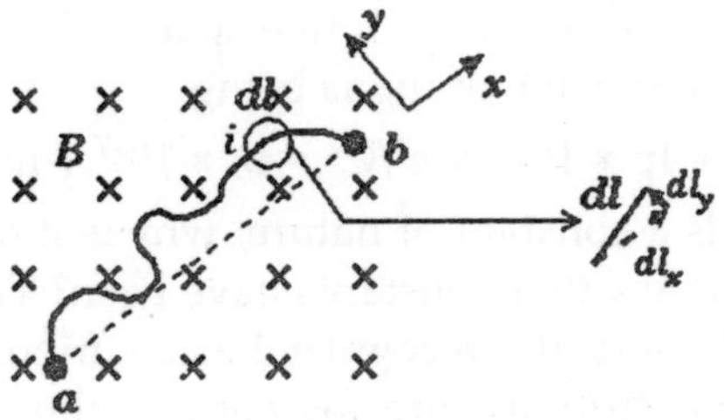

Fig. Blowup of a Section of The Wire

In breaking this distribution into infinitesimal lengths, dl, if we note that we are free to do this such that dl has components that are both perpendicular and parallel to the line between points a and b (we called these the y and x axes, respectively), then note that the contributions on the force due to the y components will cancel. In going from a to b, our net displacement is only along x. Hence, any +y displacement must eventually be countered with a -y displacement that cancels it. Since the magnetic force on any +y displacement is opposite in direction to that on any -y displacement, the integral over i^*dl_yB must be zero. Hence, we have a non-zero contribution only from the i^*dl_xB components. We note that the B field is perpendicular to both the x and y components. Hence our result is $F_{net} = iL_{ab}B$ where L_{ab} represents the distance from a to b along a straight line between the two points.

MAGNETIC FIELD OF MOVING CHARGES

In keeping with our previous statement about the association of charges acted upon by fields and the fields produced by those charges, it is only natural to think that, once it is established that magnetic fields act on *moving* electric charges, then moving electric charges should create magnetic fields. In fact, experiment shows just such a relationship! The symmetry inherent in nature between fields and charges is therefore upheld. We find that a single moving point charge q which has constant velocity v in a vacuum produces a magnetic field;

$$B = \mu_0/4\pi \times q\, v \times r/r^2$$

Note the introduction of a new fundamental constant of nature, m_0, the permeability of free space.

We *define* the value of m_0 as being

$$\mu_0 = 4p \times 10^{-7}\ N{\cdot}s^2/C^2 = 4\lambda \times 10^{-7}\ T{\cdot}m/A$$

So, if m_0 is a constant of nature, why is it defined rather than measured as other constants have been? The reason lies in what was historically recognized as a curious relationship between the permittivity and the permeability of free space. Namely, their product is the inverse of the square of the speed of light, i.e.

$$c^2 = 1/\varepsilon_0\mu_0$$

The reason for this amazing correspondence will be explained later in the semester. For now, we take it that it is useful to define the value of c and hence after measuring ε_0 we define the value of μ_0 to be consistent with the definition of c given the above correspondence.

BIOT-SAVART LAW

In determining the magnetic field for a bunch of moving charged particles which formulate a current, one must be guided by experiment. Starting from measurements of the magnetic field around a long wire carrying a steady current, two French scientists, Jean Biot and Felix Savart, worked out the relationship relating magnetic field strength to the distance from the wire to the point where the field is measured as being 1/R with R being the distance. In keeping with nature's desire for symmetry, there are no preferred directions in space, hence all points on a circle of radius R and centred on the current distribution have the same field magnitude. It was found that the direction of the field is tangent to this circle at each point. Later, it was determined that the strength of the field was proportional to the current I in the wire.

$$B = \frac{\mu_0 I}{2\pi R}$$

Starting from experimental observation of the above, a theoretical formula can be obtained that reflects the contribution due to current through infinitesimal lengths of wire all summed together vectorially through integration. This formula, first worked out by Biot and Savart, is referred to as the Biot-Savart Law is the magnetic field contribution of a current element, i.e. an infinitesimal length d*l* of current I where d*l* has the direction of I and r is the vector from the current element to the position where the field is to be calculated.

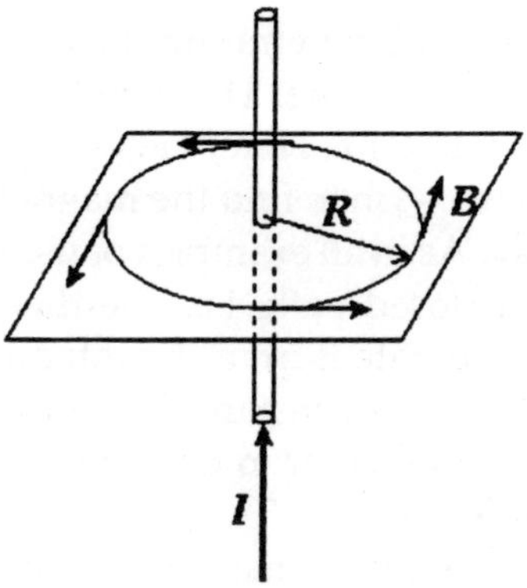

Fig. Magnetic Field Around a Straight Conductor Carrying a Current I

The Biot-Savart Law is the notion that magnetic fields obey the *principle of superposition.* This means that we can get the net field by just vectorially adding the contributions of the fields due to the infinite number of infinitesimal current elements.

Applying the Biot-Savart Law

To gain experience with applying this law, let's consider some specific examples, starting with the infinite straight wire carrying a steady current I.

DEFINITION OF AMPERE'S LAW

Andre Marie Ampere was not in agreement with Biot and Savart as to their experimental determination of the B field. First of all, isolated "current elements" such as appear in the law are difficult to justify "physically". Currents are always part of a complete circuit and hence identifying them as individual, isolated "elements" denies their true nature. In his own experimental and theoretical studies, Ampere determined an alternative law that can be shown to be in agreement with Biot and Savart's Law for the case of an infinitely long, infinitesimally thin wire.

Ampere's Law is stated as:

$$\oint \vec{B} \cdot d\vec{l} = \mu_0 I_{enc}$$

where I_{enc} is the current penetrating the area enclosed by the path over which we integrate. The obvious example to apply this to is the long, thin wire. Here, the wire is carrying current into the page and we wish to find the magnetic field at a point P a distance r away. As in all examples of using Ampere's Law, we wish to find a closed path, i.e. one that bounds an area, that allows us to separate B from the integral. To do this, we need a path in which B is constant in magnitude and parallel, antiparallel, or perpendicular to d*l* for every part of the path. The reasons for this are:

- In order to be mathematically sensible, the path must define an area through which the current passes, i.e. a non-closed path does not define a definitive area through which we can unambiguously state what current is contributing to the B field along the path. Therefore, the path must be closed.
- Parts of the path must be parallel (or anti-parallel) to the B field so that we have something to evaluate for B·d*l*. If, for part of the path, we have B perpendicular to d*l*, then, for those parts, B·d*l* = 0. In the end, we wish to know the magnitude of B. Therefore, we must have B·d*l* be known over the whole path and known in such a way that *all* of B is known, not just a component of it.
- Clearly, for this to work, you can only use Ampere's Law for cases in which you already know a lot about the behaviour of B. You need to know its direction and something about how its magnitude changes

over the possible paths you might choose. In addition, the B field must have enough symmetry to make it possible for you to complete steps 1 and 2 above.

Applying Ampere's Law to a Long, Thin Wire

The obvious path for the long, thin wire is shown below as the dashed line labeled d*l*.

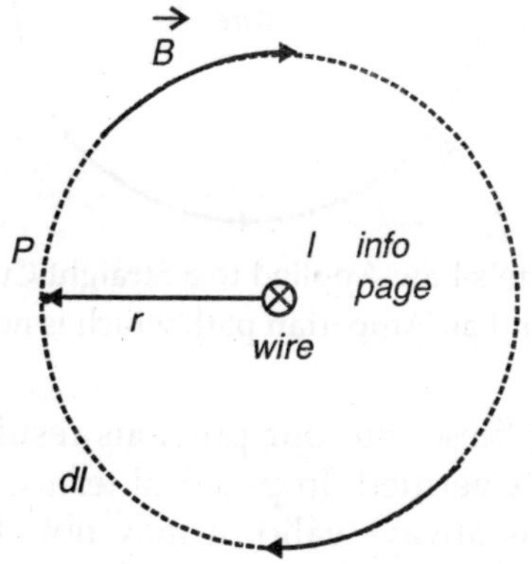

We use this circular path of radius r since we know, from the Biot-Savart Law, that the magnetic field around the wire is tangent to the circle in the clockwise direction and that it's magnitude depends only on r, not on position parallel to the wire. Thus, we use Ampere's Law to derive

$$B.dl=\mu_0 I_{enc}, \; B= \mu_0 I/2\pi r$$

To be sure that Ampere's Law is consistent with the Biot-Savart Law, let's see if it works for some path other than the circular one we chose. For example, look at the path below which has two circular arc sections centred on the wire carrying current into the page. One arc has radius a and the other radius b. They are connected by straight sections.

Note that the path is closed and that we do know the characteristics of the B field. The circled numbers indicate that we have identified 4 pieces of the path for the application of Ampere's Law. These four pieces give where we need to know what B_1 and B_3 are and we have used the fact that, for the straight paths 2 and 4, the B field is perpendicular to the path dl at every point along the path. The Biot-Savart Law also tells us that

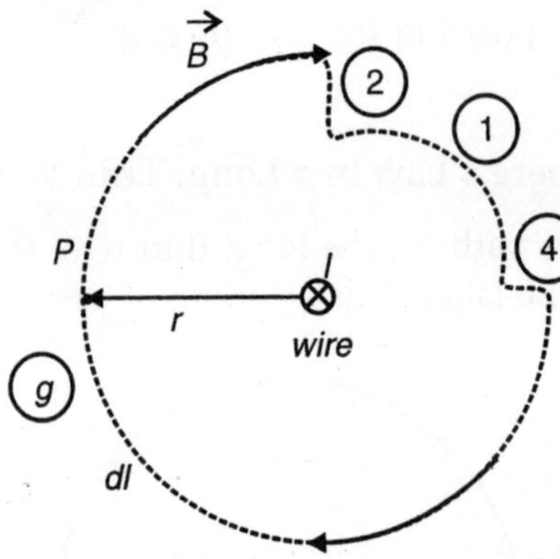

Fig. Ampere's Law Applied to a Straight Current into the Page and an Amperian path which is not Circular

$B = \mu_0 I/2\pi b$

If you plug these into our previous result, you find that Ampere's Law is verified. In practical terms, this means that Ampere's Law is always valid. It may not always be useful for finding the B field. In this respect, it is exactly like Gauss's Law in that you can only apply it to learn about the B field if you happen to recognize a symmetry of the field that allows you to bring B outside the integral of B·dl.

Magnetic Field of a Solenoid

We can readily find examples where Ampere's Law is more useful, by virtue of being easier to apply, than the Biot-Savart Law. For example, as was the case for the electric field, it was deemed to be very practical to have a device which could "store" magnetic field as a capacitor "stores" an electric field. We know that what the capacitor really stores is electric charge separation, but one of the properties for which a parallel-plate capacitor is useful is its ability to produce a uniform electric field between its plates. The equivalent device for magnetic fields is the solenoid. It's simply a conducting wire wrapped into a cylindrical shape. Even though the actual shape of the wire is helical, for densely packed wrapping we can actually consider the solenoid to be a bunch of closely spaced coils. Near the centre of each coil we know that the magnetic field is nearly perpendicular to the plane of the coil.

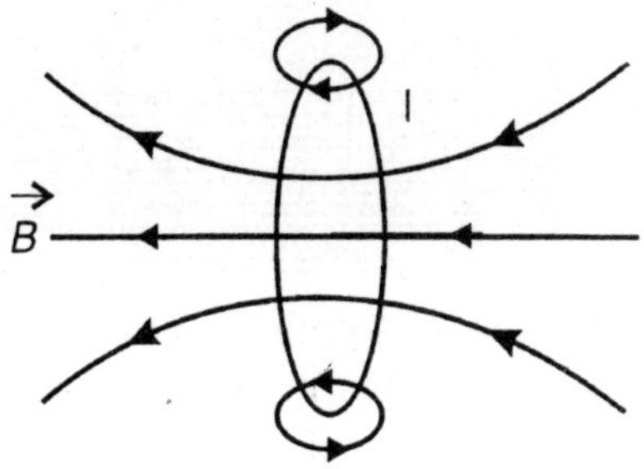

Fig. The Magnetic field near the Centre of a Single coil Carrying a Current I

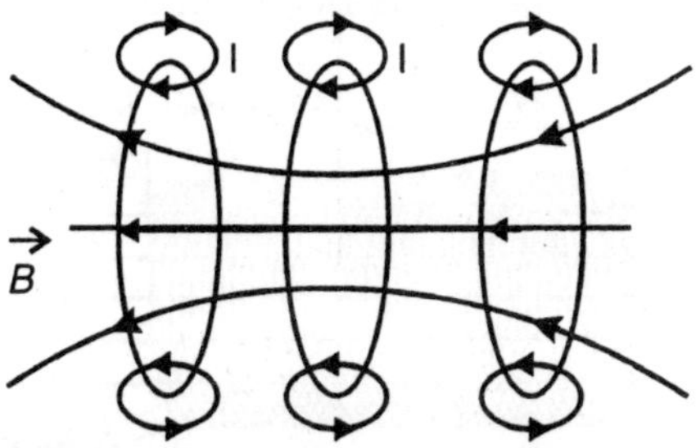

Fig. The magnetic field near the Centre of a set of three Coils all Carrying a Current I

If we place coils on either side of the first and let them all carry the same current I in the same direction, then the magnetic field lines will link together near the centre to make a field that is approximately uniform in direction and strength near the centre of the set of coils. Placing many coils in close proximity to each other yields the solenoid field.

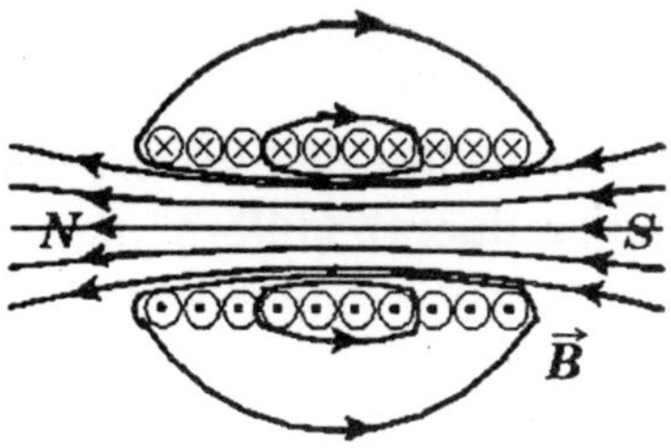

A solenoid approximates many current-carrying coils spaced much more closely than the coil diameter.

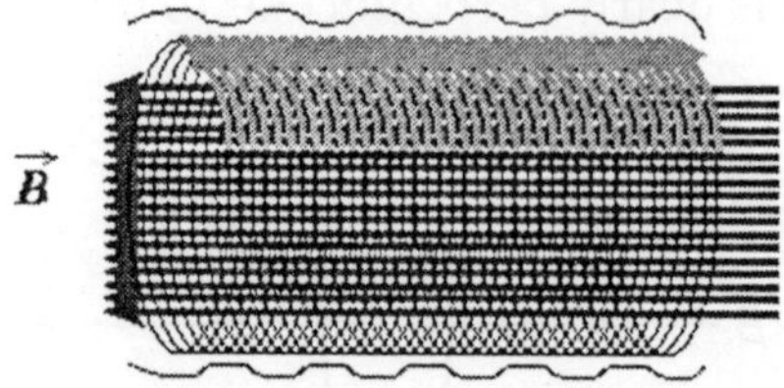

A perfect solenoid has coils so close together that the magnetic field is zero outside the solenoid and perfectly uniform inside.

Approximating the field as constant in direction and magnitude near the centre of the solenoid allows us to use Ampere's Law to calculate its magnitude.

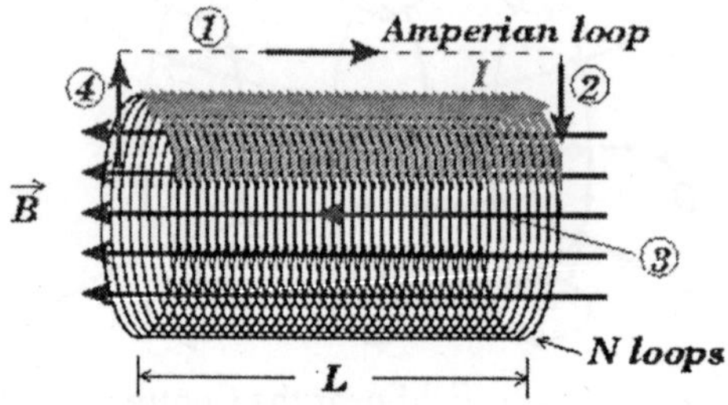

Fig. Applying Ampere's Law to a solenoid

Choosing a flat rectangle as the Amperian loop that the contributions to the loop integral can be broken into four parts. The area bounded by the flat rectangle is penetrated by N loops over a length L.Hence our result is *B.dl*= $B_{solenoid}L = m_0NI$ $B_{solenoid} = m_0nI$.where n = N/L is the linear density of current loops.

Magnetic Field of a Toroid

We can use what we just derived to examine the case of another device that produces a uniform magnetic field. In this case, we take advantage of our analysis of the solenoid to ask what happens if we bend a solenoid into a circle so that the ends join. The new configuration still has approximately zero field in the regions outside the volume contained by the coils, but the field inside that volume is again approximately uniform if the distance between the coils is small compared to the size of the coils. This device is a toroid.

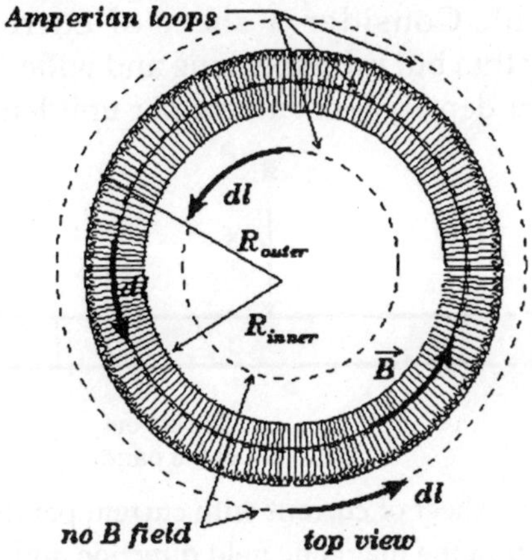

Ampere's Law applied to a toroid. Note that Amperian loops which lie wholly outside the volume contained by the toroidal coils experience no magnetic field.

We expect the magnetic field to have circular symmetry about the centre of the toroid using the same reasoning as for the solenoid. Hence, we expect it to be most useful to use circular paths for evaluating the Ampere integral of magnetic field and path. For any circular path whose area is not intersected by the coils, the magnetic field is zero and the current penetrating the area is, by definition, zero. Hence, $B_{r < Rinner} = 0$. For a circular path whose radius is greater than R_{outer}, we also expect to have no magnetic field. This might seem to be a contradiction with Ampere's Law because the current in the toroid coils definitely penetrate the area contained, however, note that the coils take current through the area (out of the page) at R_{inner} and *back into* the page at R_{outer}. So, the net current penetrating the area contained by the circular loop is zero since every loop of the wire carries current through the area twice-in opposite directions! So it is consistent that the magnetic field be zero.

Magnetic Field of an Infinite Sheet of Current

We can consider an instance of a current distribution which does not have cylindrical symmetry, but which is susceptible to Ampere's Law for finding the magnitude of the magnetic field. Consider a sheet of current which is infinitesimally thin but infinitely long and wide. The sheet has a linear current density (i.e. current per unit length), λ.

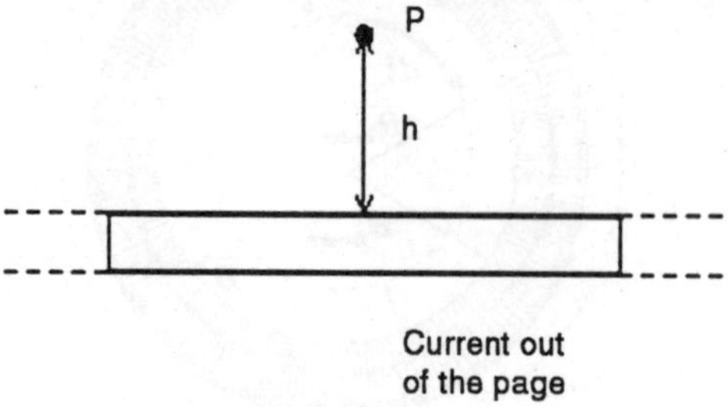

Fig. An infinite sheet of current with current per unit length λ. We wish to find the magnetic field direction and magnitude at point P a distance h away from the sheet

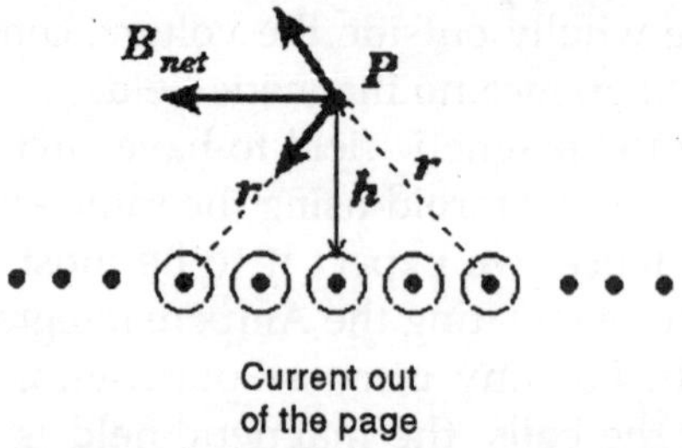

Fig. Consider a set of wires laid in place of the Current Sheet

We can replace the sheet with an infinite set of wires arranged so that each wire carries current consistent with an overall current per unit length of l as in the case of the current sheet. The wires which are equidistant from the line from the set of wires to point P have magnetic fields whose vertical components cancel and whose horizontal components add. Hence, the net magnetic field is in the horizontal direction. This field is uniform since the distribution of wires is infinite, i.e. any position for point P can be considered as the "middle" of an infinitely long current sheet. Note also that the magnetic field direction for the infinite sheet is independent of the distance of P from the sheet and that the same arguments for extending consideration of a finite set of wires to an infinite current sheet state that the magnetic field direction on the other side of the sheet has the field pointing in the opposite direction.

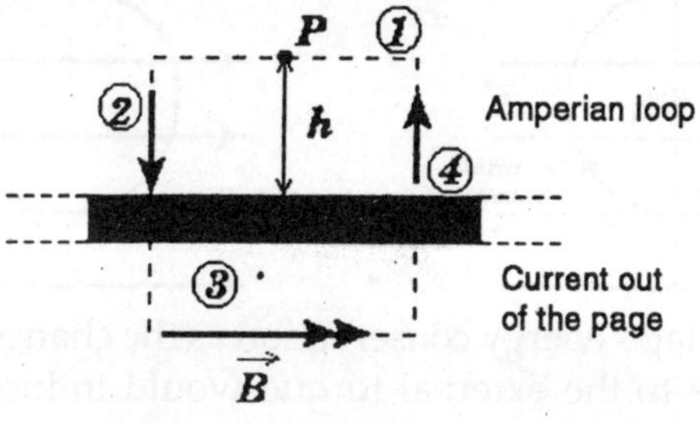

Fig. The Amperian loop for the infinite current sheet

The loop gives us $B.dl= \mu_0 I_{enc.},$ $B_1 \cdot dl + B_2 \cdot dl + B_3 \cdot dl + B_4 \cdot dl = m_0 \lambda L$. Notice that the magnetic field is always perpendicular to parts 2 and 4 of the path. Part 3 is along the field direction below the sheet since, as stated before, the same symmetry arguments work here to say that the field should be parallel to the sheet, but in the opposite direction to the field on the other side of the sheet. Therefore,

$$B_1 \cdot dl + B_2 \cdot dl + B_3 \cdot dl + B_4 \cdot dl = \mu_0 \lambda L$$

$$B_P L + 0 + B_P L + 0 = \lambda_0 \lambda L$$

$$B_P = \tfrac{1}{2} \mu_0 \lambda$$

FARADAY'S LAW

Symmetry in nature predicts that, if a magnetic field acting on a loop of conductor carrying a current can produce a torque on the loop then a torque acting on a conducting loop in a magnetic field should produce a current.

Symmetry in producing a torque for a loop with an externally maintained current going through it is balanced in nature by having an externally applied torque on a conducting loop produced a current in the opposite direction.

Note that the symmetry cannot be "perfect" in that the direction of the current that is induced in the loop due to the externally maintained torque must be opposite to the direction of current that would create a torque in that direction. This is necessary due to energy conservation. If the effect happened as in our first case then more current would go through the loop and the induced torque would increase.

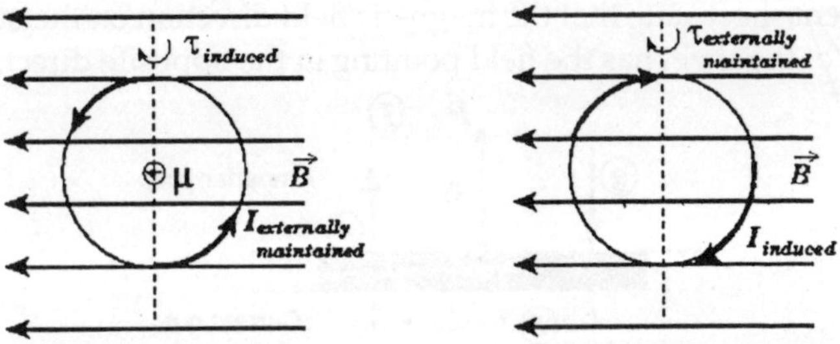

This violates energy conservation as the change in angular position due to the external torque would induce still more change in angular position. Hence we note that if the effect of induced current does take place, the current must be induced in a direction that *opposes* the external torque that causes it.

Faraday was the first to show that this induced current effect does happen. In studying the effect, we find that the magnitude and direction of the induced current effect is due to a change in the magnetic flux through an area in space. This effect is referred to as electromagnetic induction and the quantitative description of the effect is Faraday's Law. Specifically, Faraday's Law states that $\varepsilon = -d\phi_B/dt$.

In other words, Faraday expressed his effect by saying that the magnitude of the EMF induced by any change of magnetic flux is proportional to the magnitude of the change in magnetic flux. The induced EMF is what causes the current to flow. To correspond to the necessity of energy conservation, the direction of the EMF must be such that the induced current would oppose the change in the magnetic flux. This was best expressed by H.F.E. Lenz in Lenz's Law.

The Direction of any Magnetic Induction Effect must Oppose the Cause of the Effect

The easiest way to generate a static picture of this is to observe the four cases below of a bar magnet being pushed toward and away from a conducting loop. The current created by the induced EMF must oppose the change in each case.

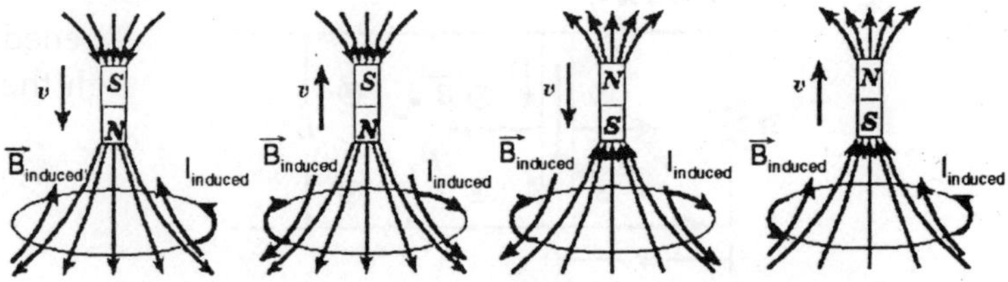

Fig. A stationary conducting loop with a bar magnet that can be alternately pushed toward or away from the loop with either of two orientations of the poles (i.e. North pole toward the loop or away from it)

It shows, in each case, how the induced magnetic field from the induced current (which comes from the induced EMF assuming the loop has some amount of resistance) opposes the change in magnetic flux.

Since magnetic flux is a scalar, it might be difficult to think of a "direction" for it, but the sign of the flux depends on the arbitrary convention for the sign for area, A. Once that is set, then the dot product of B and A gives the positive or negative sign for the flux. The negative sign in Faraday's Law then gives the appropriate sign for the direction of the induced EMF.

The induced current creates an induced field which points opposite to the external magnetic field from the bar magnet since the flux going down through the loop is increasing with time.

For the third case, the induced current creates an induced field which points opposite to the external magnetic field since the flux pointing up through the loop is increasing. Verify that the second and fourth cases shown make sense to you following the same argument. The importance of Faraday's effect is hard to overestimate. Most of the world runs on electricity generated as a result of it.

Faraday's Law Applied

If we consider the case for a conducting bar sliding over rails in the presence of a magnetic field, we have a good case for examining both Lenz's Law and considering a quantitative example of Faraday's Law.

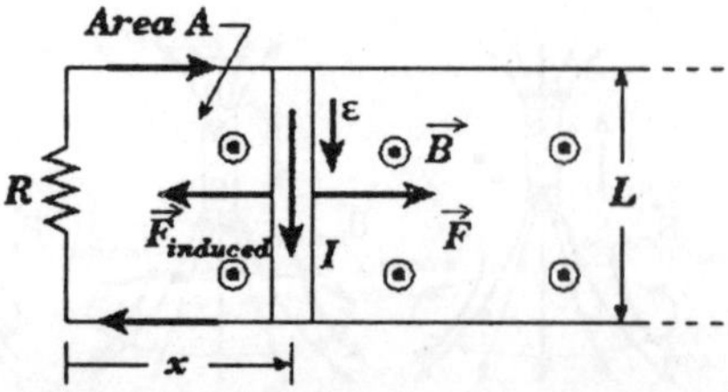

Fig. A conducting bar slides on frictionless conducting rails in the presence of a magnetic field pointing out of the page

MOTIONAL ELECTROMOTIVE FORCE

We consider conducting material to have a number (essentially infinite) number of electric charge carriers that are free to move (albeit with resistance unless the conductor is a perfect conductor) throughout the material in any way demanded by external electric and magnetic fields.

That means we can consider the source of an induced EMF by looking at the direct effects of an external magnetic field on charges in a moving conductor. For example, if we consider a conducting bar of length L moving in a plane perpendicular to a constant magnetic field.

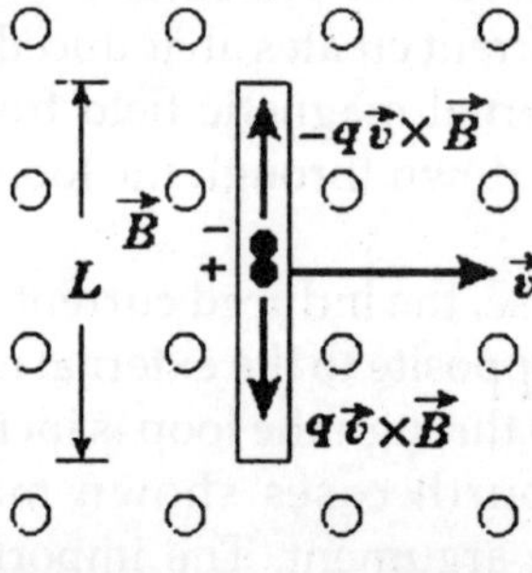

Fig. A conducting bar moving in a magnetic field experiences separation of positive and negative charges within the bar. This leads to an induced EMF across the bar

The positive charges are pushed to the bottom by the velocity cross magnetic field force and the negative charges are pushed in the opposite direction. The charge separation creates an electric field in the downward direction. This electric field builds as the charge separation grows until the force on the charges due to the resulting electric field cancels the force due to motion in the magnetic field. That means that we have, if we assume the charge q to mean positive charge, $e_{induced}$ = vBL

where the motional EMF, ε, is just the average electric field divided by the length of the bar. It is induced by the motion of the bar in the magnetic field. The magnitude of this EMF is equal in magnitude to the product of the velocity, the magnetic field magnitude, and the length of the bar. If we attach this bar to frictionless rails which provide a resistance R and provide an external force on the bar to maintain its constant velocity, then the bar is a *source* of EMF and just like any other EMF source, can drive a current through the circuit.

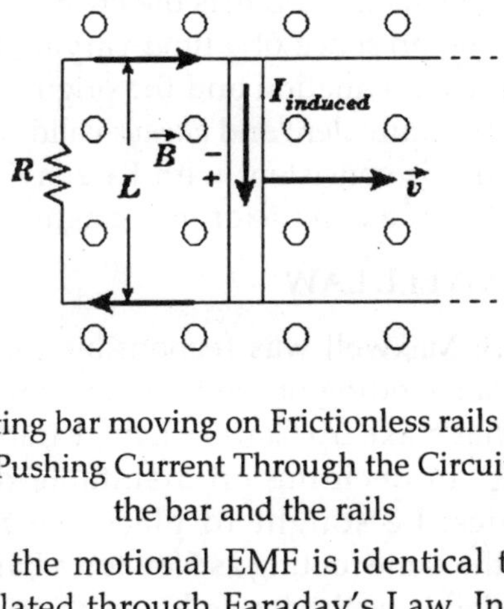

Fig. A conducting bar moving on Frictionless rails acts as an EMF Source for Pushing Current Through the Circuit formed by the bar and the rails

Note that the motional EMF is identical to the induced EMF we calculated through Faraday's Law. In fact, motional EMF is one example of Faraday's Law. It has a number of practical applications.

Note that the direction of the EMF in both the top and bottom pieces of the loop point to the right. Hence the bottom EMF pushes current counter-clockwise about the loop and the top EMF pushes current clockwise.

The motional EMF on each of the four sides of the conducting loop can be calculated. The net EMF pushes current counter-clockwise around the loop.

The net EMF for the circuit is therefore $e_{net} = e_{bot} - e_{top} = B_0 l v_y (y_{bot} - y_{top}) = B_0 l w v_y$ if we arbitrarily assign counter-clockwise as the positive direction. The induced current in the loop is $I_{induced} = e_{net}/R = B_0 l w v_y/R$.

INDUCED ELECTRIC FIELDS

In general, since the force on charge carriers in a conductor is given by F = qv × B,we can calculate the motional EMF for a conductor moving in a magnetic field via the following formulation.

$$E \cdot dl = \varepsilon \quad = \quad - dF_B/dt$$

In general, we can calculate motional EMF's, i.e. the EMF generated in *moving conductors* through the interaction of the motion and external magnetic field. The second equation shows that Faraday's Law predicts the creation of an induced electric field in the presence of a time-varying magnetic flux. Note that this second equation and the original expression of Faraday's Law are equivalent and *always* valid. The calculation of motional EMF's is consistent with Faraday's Law *only* for the case of motion of a conductor in a magnetic field.

AMPERE-MAXWELL LAW

James Clerk Maxwell was responsible for setting down our current understanding of the fundamental importance of the equations that make up Gauss's Law, Faraday's Law, and Ampere's Law. In deciding on a suitable description of electrodynamics, he sought to place the mathematical description on the same footing as Newton's Laws of Motions. That he succeeded is testimony to his mathematical skill and great physical insight. To start, we should review here the form of the laws as we currently describe them.

Gauss's Law for Static Electric Charges	$E \cdot dA = q/\varepsilon_0$
Gauss's Law for Magnetism	$B \cdot dA = 0$
Ampere's Law	$B \cdot dl = \mu_0 I_{enc.}$
Faraday's Law	$E \cdot dl = -d\phi_B/dt$

where the second equation represents the fact that there

are no magnetic charges observed (so far) in the universe, hence there are no point-like sources or sinks of the magnetic field. We first note the degree of mathematical symmetry present in the equations despite the differences in behaviour of electric and magnetic fields.

It's also clear that this symmetry between electric and magnetic fields would be perfect except for two obvious asymmetries within the equations themselves. The first is that there are no magnetic charges as there are electric charges (something that has motivated physicists to seek out such charges for decades) and the second is that, electric fields can be created in *two* ways: either from static electric charges (Gauss's Law) or induced by time-varying magnetic fields (magnetic fluxes to be exact), but there is no corresponding possibility of creating magnetic fields from either static magnetic charges (they have not been found to exist) nor from time-varying electric fields.

If we simply assumed that the latter symmetry had to exist, why would it not be apparent in experiments that were done up to the 1800's? In this, we can make a small digression that points out one of the prime means of gaining insights into physics research.Let us infer that nature is respectful of symmetry mathematically. What would Maxwell's Equations look like completely symmetrically?

First, the existence of magnetic charges would change the form of Gauss's Law applied to static charges. We would have $E \cdot dA = q_E/\varepsilon_{0,}$ $EB \cdot dA = \mu_0\, q_{B..}$ If units are to be kept consistent. What about currents involving electric and magnetic charges? Certainly there would be contributions to both Ampere's Law and Faraday's Law. These contributions, again looking at consistent units, would look as follows: $E{\cdot}dl = -d\phi_B/dt\ \mu_0 I_{B.}$ where the subscripts E and B depict electric or magnetic charges.

The k in the revamped Ampere's Law is a placeholder for the correct proportionality constant for this term. We can find the value of this constant in terms of the fundamental constants m_0 and e_0 purely from dimensional analysis. Since magnetic monopoles do not exist, we deal simply with the additions necessary to consider induced magnetic fields from time-varying electric field fluxes. This would affect only Ampere's Law and the term added would assume that a time-varying electric flux would induce a B field. To make sure that the units are correct, we would want to have Ampere's Law extended as follows:

$B\cdot dl = \mu_0(I_{enc.} + I_{displacement})$,

where we want $I_{displacement} = d\phi_E/dt$,

That is, this new term $I_{displacement}$ is proportional to the time-rate of change of electric flux. The simplest form of converting electric flux (SI units of $N/(C \cdot m^2 \cdot s)$) to current (C/s) is, interestingly enough, multiplication by a constant with the units $C^2/(N \cdot m^2)$. Note that these are precisely the units of the permittivity constant, ε_0. Hence, $I_{displacment} = \varepsilon_0\, d\Phi_E/dt$.

This comes directly from dimensional analysis and a "belief" in the need for symmetry among the equations for electricity and magnetism. Note that this ansatz also gives rise to an assumption as to why the effect might be hard to observe. Note that, if the above form is correct, then Ampere's Law is changed to have the form $B \cdot dl = \mu_0 I_{enc.} + \mu_0\varepsilon_0\, d\Phi_E/dt$.

This means that the time-varying electric flux term is multiplied by *two* small terms rather than just one. This might make the effect weak enough to be masked by other experimental interferences unless one is very careful.

In fact, the form we have derived for $I_{displacement}$, referred to as the displacement current, does exist and can be verified experimentally. It was motivated by Maxwell theoretically not from the symmetry and dimensional analysis argument above, but from the following consideration of a weakness in Ampere's mathematical description of his law.

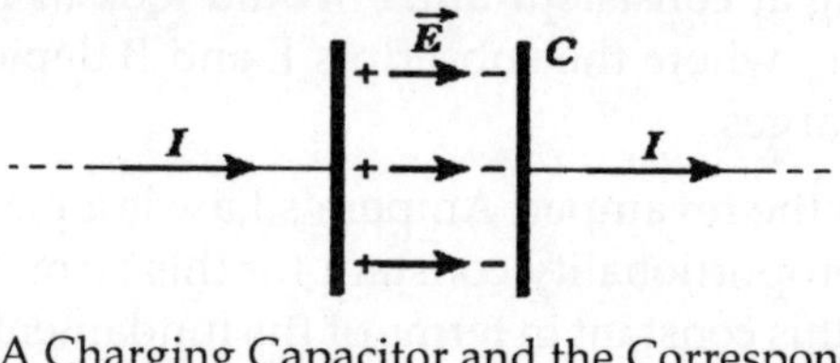

Fig. A Charging Capacitor and the Corresponding time-varying Electric field between its Plates

Consider the case of a capacitor connected to conducting lines. If one of the lines carries a current I, then the other line connected to the other plate must also carry such a current due to the electric field of the previous plate.

If we apply Ampere's Law to a circular disk around the wire, then the magnetic field is expected to be constant in magnitude and to have a direction which follows the boundary. The magnitude of the B field is proportional to the current penetrating the area bound by the path.

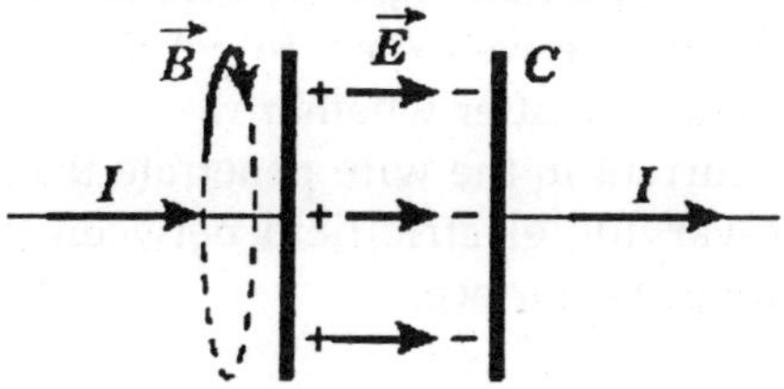

Fig. Magnetic field Around a current as it Approaches the Capacitor

Unfortunately, the mathematical description of the area bounded by a curve is rather loose.

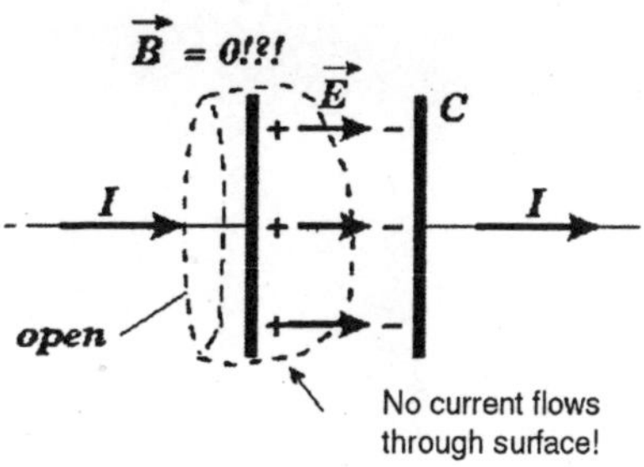

Fig. Magnetic field around the same perimeter which encloses a different area

Ampere's Law, as stated, is a line integral which can be shown to be equivalent for the two areas and in fact for any area bounded by the same circular curve around the wire in the region outside the capacitor.The unfortunate part, though, is that no current penetrates the second surface and hence the prediction is that the magnetic field should be zero around the boundary.To resolve the obvious inconsistency, Maxwell generalized Ampere's Law to include the displacement current matching just the characteristics we gave on the basis of the dimensional argument above.

To verify the form of the displacement current, note that a parallel-plate capacitor builds charge according to the current impinging on the left plate for our figure. This current is related to the change in electric flux as follows: $\Phi_E/dt = I/\varepsilon_{0.}$ If we multiply this by e_0, we see that we get the displacement current as being equal to the current in the wires attached to the plates of the capacitors, hence we expect to get the same value for the magnetic field no matter whether we draw the surface so as to have the current in the wire penetrate the surface or to have the time-varying electric field between the capacitor plates penetrating the surface.

Index